GOLD PANNING
Colorado

A Guide to the State's Best Sites for Gold

GARRET ROMAINE

GUILFORD, CONNECTICUT

To my best old buddy, Val Bailey

FALCONGUIDES®

An imprint of The Rowman & Littlefield Publishing Group, Inc.
4501 Forbes Blvd., Ste. 200
Lanham, MD 20706
www.rowman.com

Falcon, FalconGuides, and Make Adventure Your Story are registered trademarks of The Rowman & Littlefield Publishing Group, Inc.

Distributed by NATIONAL BOOK NETWORK

Copyright © 2018 The Rowman & Littlefield Publishing Group, Inc.

TOPO! Maps copyright © 2018 National Geographic Partners, LLC. All Rights Reserved.
Maps © Rowman & Littlefield
All photos by Garret Romaine unless otherwise noted

All rights reserved. No part of this book may be reproduced or transmitted in any form by any means, electronic or mechanical, including photocopying and recording, or by any information storage and retrieval system, except as may be expressly permitted in writing from the publisher.

British Library Cataloguing-in-Publication Information available

Library of Congress Cataloging-in-Publication data available

ISBN 978-1-4930-2856-6 (paperback)
ISBN 978-1-4930-2857-3 (e-book)

∞™ The paper used in this publication meets the minimum requirements of American National Standard for Information Sciences—Permanence of Paper for Printed Library Materials, ANSI/NISO Z39.48-1992.

The author and Rowman & Littlefield assume no liability for accidents happening to, or injuries sustained by, readers who engage in the activities described in this book.

Printed in the United States of America

CONTENTS

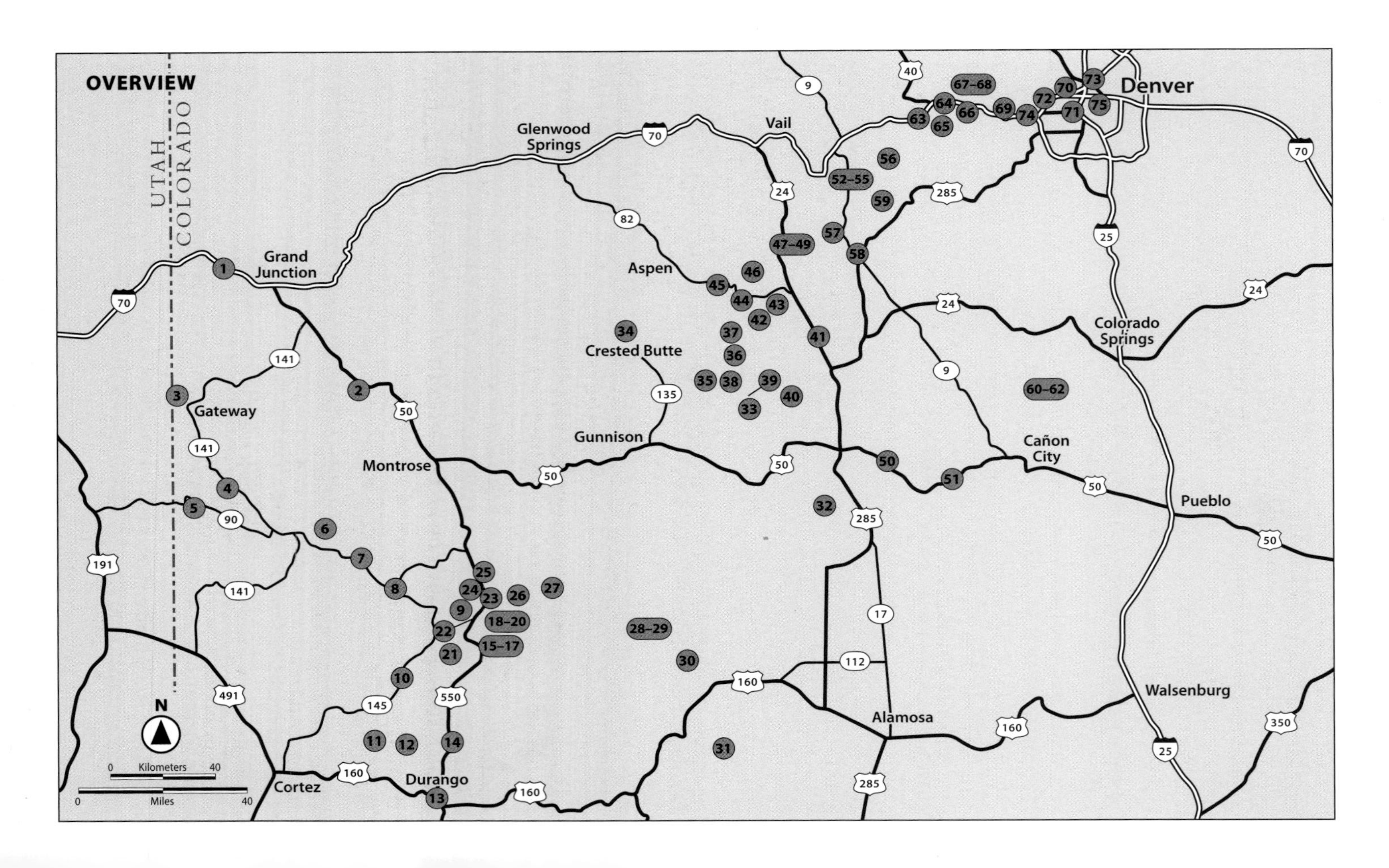
OVERVIEW
UTAH
COLORADO
Denver
Glenwood Springs
Vail
Grand Junction
Aspen
Crested Butte
Gateway
Montrose
Gunnison
Colorado Springs
Cañon City
Pueblo
Walsenburg
Alamosa
Cortez
Durango
N
Kilometers
Miles

ACKNOWLEDGMENTS

I first dipped a pan in Colorado gravels in the 1990s, and I've come to greatly enjoy the time I've spent scouting around this great state. It has been a fascinating journey to far-flung old camps and abandoned mines and a true test for my Jeep, affectionately named "Cherry Blossom."

Here's a genuine "recon salute" to the many folks who helped me out in the field, but especially two great mining buddies: Frank Higgins and Dirk Williams. They're good sports about going on frantic recon patrols, but they would much prefer to hunker down in a single spot for an entire week. Well, this time we got to. Here's a salute to patient road trip companions, great campers, and strong backs when it comes to shoveling.

Special thanks go to my long-suffering wife, Cindy Romaine, who consented to my extended absence. She's now officially up for sainthood.

Geology nerds ready to go underground. From left, Dirk Williams, Garret Romaine, and Frank Higgins.

PREFACE

This book is aimed at everyone who has had even a minor case of gold fever. If you have ever seen a small pinpoint of gold emerge from the black sands and concentrates at the bottom of a gold pan, you know what I mean. If you have never experienced the thrill of recovering an elusive piece of gold, you have a major emotion in front of you. There are very few things that quicken the pulse like seeing actual gold gleaming in your pan.

I have panned gold in the West's historic mining districts since the 1970s, primarily as a columnist for *Gold Prospectors Association of America*, the magazine of the Gold Prospectors Association of America (GPAA), and before that as a senior editor at *North American Gold Mining Industry News*. I have learned where to go, places to avoid, and what to find. I cannot guarantee to make you rich, because this is not a get-rich scheme. It's a labor of love, a lot of fun, and a ticket to some of the most interesting places out in the hills. What I can do is save you time, effort, and money as you research places to explore, and introduce you to the thrill of driving from the bottom of a drainage all the way to the top, and that last corner you turn when an old, wooden structure comes into sight, bronzed deep brown from the years of sunlight and preserved in the thin, dry mountain air.

Hiking to abandoned mines is another adventure I wish I could share—there is a true sense of accomplishment when you reach the end of a trail and find that old mine shaft. Sometimes you can close your eyes and imagine hundreds, if not thousands, of noisy miners and merchants in those old settings, only to be interrupted by the call of a stern old turkey. I have watched many old mills and headframes crumble into masses of rotted timbers, but the nearby tailings piles still yield great hand specimens. Every spring, fresh torrents move and mix the sands and gravels anew and re-sort the pay streaks. There is still a lot of gold out there, if you have the time and energy to go about recovering it. Most of the easy stuff is gone, but that just makes the thrill of finding what's there that much greater.

By reading this book, you will save yourself a lot of trouble out in the wilder parts of the region, and you will benefit from some of my worst mistakes. I have seen gates spring up across once-reliable access roads, but I've watched GPS and Google Earth become major forces for preventing

confusion in the field—while adding some of their own. I have collected hundreds of GPS coordinates for you to use as aids, but you may find a few of your own, too. Whether you have an old gold pan under the seat of your truck or a new Keene sluice still in the box, you will benefit from this book.

I'm going to err on the side of providing you with as much geology as I can. Where possible, I will provide explanations from some of the state's great geologists about host rocks that contain quartz veins, and will try to explain the associated minerals that brought so much prosperity to this region. I will try to educate you about collecting rock samples, and point you toward the tailings piles that yield some good specimens. I will follow that up in appendix A by discussing the common modern machines that grizzled old prospectors would have killed for back in the day. By the end of this book, you'll find gold, you'll learn some history, and you'll have a great appreciation for the gold rush days that settled this region. Your collection of photographs will be a source of pride. Most of all, you will visit some of the storied districts that gave Colorado much of its resilience and vigor. You'll visit places that you'll want to talk about with your friends and family, and you'll see wildlife, landforms, and vistas that will stir your soul. I hope you enjoy your search.

Colorado is a big state, with a growing population. Environmental concerns, land restrictions, and private property are a continual issue for fieldwork, and that will only get worse. Many historical societies are putting a lot of effort into preserving old buildings and artifacts, and you should never interfere with their efforts. Donate if you can—they are doing great work. Also, when you see a local museum, drop in and check it out. Chat up the staff and be ready to purchase a book or two. They need your support. Along the same lines, whenever you see a rock shop or a place selling prospecting supplies, hit them up, too. Purchase something and ask about local conditions. They want you to be successful just as much as they want to stay in business, so check them out.

For this book, I've made a dedicated search of every promising panning locale in Colorado, thanks to hundreds of hours of research. I verified access, GPS coordinates, and colors. Just for kicks, we panned the floor mats and wheel wells from the Jeep every once in a while, and sure enough, she was collecting tiny colors for us. A lot of these areas have been hit hard over the years, but don't let that stop you. Always be on the lookout for claim markers, but you may find there are less claims going forward. I hope that's the case, because it will free you up to do more exploring with less worry.

INTRODUCTION

Of all the legends and stories about gold, the one that is hardest to explain is the adage "Gold is where you find it." The old saying seems to imply that if you look hard enough just about anywhere, you can stumble upon a nugget or two. Nothing could be further from the truth. A better way to modernize this would be to point out that "Gold is where they found it." The reason is pretty simple—the gold prospectors that came before us were really, really good at what they did. For all the images conjured up of a slightly addled Yosemite Sam with his burro and pick, there have been a lot of smart geologists since then that scientifically sampled every major gold district. Many mines started production, exhausted sizable ore deposits, and then disappeared, but there is still color in the nearby streams. The likelihood of discovering a new district is extremely remote, but they didn't get it all, by any means.

The good news is that while they were efficient, the original gold rush miners were also in a hurry. They creamed off all the easy gold—the concentrated pockets, the big nuggets, and the "potato diggings" monsters went to the smelter early. Their primitive equipment skipped a lot of fine gold. Some machines simply washed the "flour" gold over the transom, out the end of the dredge, to remix with the rest of the tailings. Miners later resorted to mercury in order to collect some of the fines, but very few miners use mercury anymore. It is toxic, poisonous, and dangerous, and modern equipment can do just as well, if not better, at recovering fine gold—which is good, because that's a lot of what's left. If there is one scientific fact you can count on, it is that it's always going to be impossible to get all of the gold from any particular area. There will always be a little color left, deep in the crevices, under the big boulders, or lying on hard clay layers known as false bedrock.

A dedicated weekend hobbyist, recreational prospector, or otherwise-motivated gold miner can still go one of two routes in the search for gold:

Hard-rock mining involves dynamite, cyanide leaching, crushers, big rigs, and lots of capital. There are still fortunes to be had out there with hard-rock mining, but the odds are long. Throughout most of the West, any sizable deposit has likely been identified and claimed up. You could still get lucky, but it will take a lot of work. You could rework old tailings and do well, or you could find an area thought to yield copper that has good gold left. Hard-rock mining is capital intensive, and it requires a lot of big machines, but if you can return from your trips with interesting rocks, you could still get lucky finding a lost ledge high up on the hill.

Placer mining involves equipment ranging from a gold pan to a boat-size dredge or wash plant. This type of gold production usually involves less investment, and will consistently yield small amounts of gold, with occasional bonanzas for those who are persistent. If you can learn to reliably return from every trip with decent concentrates, so that over time you fill a five-gallon bucket, and then maybe even a fifty-five-gallon drum, you'll do well. Your black sands, magnetite, ilmenite, rare-earth elements, and gold will be a nice reward in the long run.

Either way, your long-term goals are your own. Very few prospectors are simply in it for the money, looking on this as a way to become a millionaire overnight. Some gold panners just like to get out of town, camp in the mountains, and enjoy the spirit of the outdoors. Some people like to work up a little sweat and appetite, improve their health, and learn a little. Some of us like to solve problems and run machinery, and enjoy the challenge of making sure the equipment is running right. Still others like the wildlife, the scenery, the photography, and the historical importance of the Wild West, and bring back their riches primarily as photos and videos. In each case, if you toss in a little gold fever as motivation and stay scientific about your sampling and exploration, you will prosper far and above the value of your recovered material.

Still, a nice payday is always a treat. One sure way to reach that goal is to keep trying. Keep practicing, keep exploring, and keep getting out in the field. Another truism that seems to hold is that the farther away from civilization you get, the better your chances. Be prepared to hike a little farther from where you park than the last person.

GETTING STARTED

The best way to use this guide is to treat it like a general reference for the state. This is a guide mostly for beginners and families, so the mine tours and public panning areas are prominently featured. I tried to concentrate on the best tours and museums in the state, as those are the best bets for family fun and a good introduction to what the state has to offer, but even experienced gold panners will gain valuable insights into historic districts worth visiting.

Here are some likely scenarios:

1. You might live near a district listed here or have always wanted to visit a site. Based on what you read and learn, you will have more information to process and then decide if you want to explore that area further. Thanks to the GPS coordinates found in these pages, you can easily pinpoint a few spots that you will be able to drive right to, and thus maximize your time in the field so that you are efficient and productive.

2. If you already have a business trip, family vacation, or other reason for driving through an area, you might check these pages and determine that you will be close to a spot that is worth exploring. The information here can help you minimize the detour, making it easier to justify. In addition, if you need to camp, I have listed developed and primitive camping options where possible.

If you are considering a major investment in mining supplies, I have some good information for you in appendix A. Over the course of writing nearly one hundred articles for *Gold Prospectors* magazine, I have reviewed many products. I will provide you with good, practical information about metal detectors, dredges, high-bankers, sluices, and more. Note that I stuck that information at the back of the book so that you can get right to the good stuff: where to go.

If you are just getting started in the hobby, I recommend picking up *The Modern Rockhounding and Prospecting Handbook*, a companion book I wrote for FalconGuides. The information there can make you a better field geologist by explaining the basics of economic ore deposits, field sampling, and identifying hand specimens. If you need help identifying rocks and minerals, I wrote a rock-and-gem identification book as well: *Rocks, Gems, and Minerals of the Rocky Mountains*.

In addition to the GPS coordinates for these locales, I have also included road directions. I have to warn you though: Conditions change fast. US Forest Service (USFS) or Bureau of Land Management (BLM) offices may schedule major roadwork right over your prized vacation window. July is a particularly harrowing month; the days are long and hot, but afternoon thunderstorms can be fierce. One resident told me there are only two seasons in Colorado: ten months of winter and two months of relatives.

Washouts, fires, and other road damage can leave you sputtering in front of a hand-lettered sign at two in the morning, facing a four-hour detour (if one is even possible). Colorado's mountain passes can be fickle—summer is the only time of year when crews can get out and fix problems, and they may have to close the pass to do it. It is important that you take the road directions I provide with a healthy dose of caution and either phone the local government agency managing that land or consult their website if you don't have time to track down a human. The more you are counting on taking your family and friends to a single location for a long stay, the more important it is that you check in ahead of time.

Mountain Driving

The last few miles to the top of a drainage can be harrowing. The roads often worsen and narrow considerably, and even with the requisite four-wheel drive, you still may be creeping along at 1 mile per hour. Creek crossings can become a major challenge, too. Make sure you have the right vehicle, and don't push it. A couple I met in Telluride told me that they love their Jeep too much to take it on the mountain roads, so they rent one and relax while they conquer the passes.

Some of the high mountain roads in this book are essentially one-way as they get close to the top or near the end. That means you have to always remember where the last turnout was—especially if you're coming down the hill. The rule is that drivers coming up the mountain have right-of-way, because it's really easy to lose control when backing up downhill. Use caution, turn on your lights, honk on blind curves, keep an eye out for oncoming traffic, and remember those pullouts. You'll learn a lot about a fellow driver on a mountain road when you inevitably get to a spot where someone has to back up. It's no fun being bumper-to-bumper, trying to figure out who has to yield. One trick when you're in an area such as Telluride or Animas Forks, where the roads are crawling with rigs, is to get into a convoy. You'll have more strength in numbers, and the driver at the front of the line will lead the team.

If you do rent a Jeep in town, try to remember the "Jeep wave" when you spot another Jeep coming at you. It can be as simple as a finger lifted from the steering wheel, or a full hand wave if you're feeling enthusiastic. Jeep drivers can be very tribal, and they like to acknowledge other Jeeps as they drive by. The Jeep wave roughly translates to "Nice Jeep," "Howdy," or "Thanks for not hitting me." I sometimes find myself inadvertently giving a wave to anyone I see out on some of Colorado's lonely back roads, just to acknowledge a fellow traveler.

HOW TO PAN FOR GOLD

Since this is a guide mostly intended for beginners, I wanted to go over some of the key concepts you should understand when you go out and gather your first sample and pan your first pan.

Materials Needed

- Shovel or trowel to load in material
- Gold pan
- Snuffer bottle to capture your concentrates
- Screens (optional)
- Location, location, location . . .

Technical writers trying to help you install your new router or hard drive will always tell you to read the instructions first, all the way through, before you start. I'm guessing that you'll actually do that here, because not many of you are going to bring your book down to the creek and follow the instructions step by step while your feet are in the water. So read through these steps and *imagine* that you're on the banks of the Arkansas River, then see if it all comes together for you when you put the steps into action.

Or, if you're practicing at home, you can fill up a tub of water for your panning station. A good plastic tub would be about 2 feet wide and 3 feet long, filled with about 9 to 12 inches of water, minimum. A tub that brick masons use to mix mortar works very well, and it has nice rounded corners so it's easy to clean. The more water you have, the less problems you'll face later with muddy water that is hard to see through. Some folks use a running hose to keep the water in their tub clean, so consider that. But if you're only running a couple pans of store-bought concentrates, you should be fine with the water you start with. If you brought home some concentrates from an outing, chances are good that the sample is also very clean and won't muddy things up.

If you're out at a creek or river, you need to find a place where the water isn't rushing through too fast. You could lose your material—or even worse, lose your pan—if you get out into a fast current. You'd be surprised how far pans travel under the water, too. The team lost one once on a very small creek and it washed downstream about 50 feet.

If you don't have rubber boots on, you may want to stack a few big rocks up so that you can stand on them safely and keep your feet dry. During the summer months, on many Colorado waterways it's easy to just stand in the water in sandals or even barefoot. Some of these waters run really cold all year, however, so just do what you have to do. Panning in your brand-new Nikes is always going to be a challenge, so plan accordingly.

Any gold pan will do, from the cheaper black models about 9 inches across to the bigger 16-inch green or blue monsters with two sets of riffles. For this exercise, assume that you have one of the nice blue Keene models, about 12 inches in size, with coarse and fine riffles. If you don't have one yet, get it the next chance you have. They're worth it, and consider how the original forty-niners would have marveled at it. According to legend, the Chinese were the first to hammer one side of their metal pans into riffles, and the idea has stayed with us ever since.

Where to Get Your Sample

As you look out over the creek or river, take a few moments to scan the area for obvious catches or other features. The one thing you don't want to do is simply start digging where it's easy. First, a lesson I learned way back in 1980. My wife signed us up for a one-day extension course on gold panning at the local university, so we bundled up and headed for a spot the guide knew well. When the instructor turned us loose, I immediately grabbed a quick sample from the sandy banks, and the instructor broke out into a big smile. "Whoa, there. You want to get a better sample than that!" He then patiently walked me through some of the following things to look for:

Gravel bars: Is there a gravel bar forming? Whenever a creek or river bends, it usually starts to form a gravel bar downstream from the bend. At the top of the bend, the heaviest material falls out first, while the lighter material moves on downstream. So your best option is to look for one of these gravel bars and move to the very top of the bar, then start a hole. You'll have to move a lot of rocks out of the way, which is when a screen comes in handy.

Islands: Is there an island forming in the middle of the creek? The very top of that bar is where the heavy material drops out. That's why fish spawn at the bottom of an island or bar, because the water is calmer and it won't wash away their eggs. You want the top of an island, where the rocks first trap material that comes down and hold it in place while the water washes away the lighter material.

Past flooding: Is there any evidence of flooding from years past? Every year during spring runoff, the water level rises quite a bit. You may be able to look around and notice leaves and debris trapped in trees or bushes far above your head. Floods are a good thing when it comes to moving gold around. As you look out at the water in the summer, the only place where gold could possibly be moving is at the very bottom of the channel, at the tip of the V. Everywhere else, the gold is stuck and won't likely move. It helps if you can imagine what the channel looks like when the water level is 20 feet above where it is in the summer, because you may spot dried-up runs or abandoned paths. When the gold stopped moving as the flood subsided, it dropped out in these small channels. If they are near the top of a gravel bar, so much the better.

Bedrock: Is there any bedrock showing? Bedrock is the rock that won't move. All other gravel, sand, and mud sits on top of the bedrock. In many places there can be as much as 20 or more feet of overburden, and you won't see anything except for soil and plants, with a few big rocks. That's usually not a good place to pan, although you can still get the fine "flood gold" that moves during flood events and sprinkles the top of the riverbank. During earthquakes, heavy material such as black sands and gold will settle through the sand and mud, to eventually rest on the bedrock. During floods, when material is moving quickly, the water may scour all loose material away. At those times the gold and black sands will move across the bedrock, but any crack, crevice, or hole will trap the heavies.

Big rocks: Are there any big rocks that you can move? The old-timers often said that the biggest gold was found among the biggest rocks. The idea makes perfect sense—it takes a lot of force to move large pieces of gold, and when those big rocks are moving around, that's a perfect time to move big gold, too. Big rocks act as a natural riffle in the water flow, and they usually form a kind of "beard" behind them. This is where heavy material such as gold and black sands will settle. The water speed behind the big rock is slower than in the main channel, and gold will slam into the rock and work its way against or underneath that trap, then drop out because the water velocity isn't fast enough to keep the gold moving. That means that if the water level has gone way down and the rock or boulder is now high and dry, your job is to push it aside and dig out the material beneath it and slightly downstream.

Crevices: Are there any obvious traps, cracks, or crevices? If you think of the adage "Work smarter, not harder," you will come to love those cracks and crevices. There are several ways to dig them out—you may even see someone with a very fine dental pick digging around in a crack. Miners in the early days of the California Gold Rush used steak knives to pry big nuggets out of obvious cracks, but those days are long gone. You can use a pointed geology pick, a hammer and chisels, a whisk broom, a toothbrush, or a turkey baster and alternately scoop, scrape, or slurp up material from a crack. Your attack will vary depending on whether it is wet or dry. Some folks even use a battery-powered hand vacuum to liberate material from a crack. Others break them open even farther with sledgehammers. If you start to find big gold in a crack, you'll know what to do.

Moss: Is there any moss on the boulders, or small grasses growing from cracks in the rock? In many parts of Colorado, you don't want to start removing any vegetation, but if you're not on a Wild and Scenic River or inside a park or campground, moss mining is a known way to get yourself a good sample to pan down. As you scrape moss off a boulder, you'll see lots of material still clinging to the rock, and you can sweep it off or wash it off. Try putting the moss into a bucket full of water and squeezing, scraping, shredding, and otherwise breaking down the material. You may be surprised to see black sands clinging to the roots or moss after several washings, but when you're finally convinced you got it all, you can easily pour the bucket into a pan and prepare a sample that way. In some mining camps it was popular to dry out moss and burn it in a large metal pan, then pan out the ashes.

Cobbles: Are there smaller rocks, maybe fist-size or the size of a soccer ball, that you can clean? One good way to get a rich sample is to fill a bucket with water and systematically wash off the muddy or dirty rocks. Gold has a habit of sticking to rocks such as these, and you don't want to casually toss them aside. If you see a small run where the creek formed a new, temporary channel during flood stage, leaving behind a trail of cobbles when the water level retreated, this could be a good place to wash some rocks. You can use an old dishwashing brush, for example, or an old paintbrush. If you have a lot of smaller rocks, you can set a screen over the top of the bucket, pile on the rocks, and use a second bucket to keep pouring water over the material, then dump the tailings into a pile.

Prior work: Are there signs that someone else has been digging there? Colorado is a big state, and there aren't that many secrets about where to go.

Gold, after all, is where other people have found it. So if you're at a popular place listed in this book, you may see signs where someone has been working. There are basically two kinds of "diggings": a shallow, haphazard hole that didn't go very far, and a second, serious hole that moved a lot of material. The first hole is probably from someone in a hurry, who gave up quickly because they didn't find anything. The second hole shows someone encouraged by early results to keep going. If you find that second hole, take a few minutes to dig it back out from any material that fell in after they left, and see if you can reopen the deepest part where they probably got the best sample. Generally these holes can be very helpful, and you'd think it would be the polite thing to do to not fill them in, out of courtesy to the next person who comes along. But the simple fact is that an area can quickly acquire the look of a war zone, and it usually takes very little time to fill your hole back in.

Hopefully, you answered "yes" to at least one of these questions. If you have all of these different features in one place, you are definitely in a good spot to sample. Plan out your attack accordingly, and if you have the time, get samples from each type of deposit. Use a simple Rule of Threes—try three pans each from three different places. Once you've panned out nine pans of material, you'll have a very good idea of what you're working with. Don't throw away your concentrates! Collect them and take them home. You never know how much micron gold you have until you look at it with a hand lens or loupe.

The more concise you are with your sampling, the better you'll learn a spot quickly, so one technique is to use three different buckets, gathering them all together before you start the panning steps.

The Basics of Panning

So let's talk next about how to pan down a sample. In a nutshell, these are the steps:

1. Load
2. Liquefy
3. Stratify
4. Process
5. Inspect

Let's go through the process in more detail:

1. Pile up your pan with material. You may feel greedy and want as much in there as the pan can hold, but try not to do that when you're

first starting out. Fill the pan about three-quarters full so you can get water in and move it around. For this exercise, it doesn't matter if you screened the material down or not. Let's assume you didn't. Screening can be a two-edged sword: On the one hand, you easily get rid of a lot of extra material that has only a very slight chance of containing gold, and you can move a lot more material quickly. It takes some work to screen your pay dirt, and it's best if you have one screen with big holes and one that's much finer. But the trade-off is that if you are in a spot with big, chunky gold, you could accidentally toss it aside.

For your first sample, you might want to just use the coarsest screen. That way you'll have fewer big rocks in your pan, and at the same time you won't run the risk of losing good gold. As you keep panning your area, you might find there are no big flakes, and you can safely screen material until you start seeing bigger pieces. But it's a good idea to be systematic about it, and check yourself with a completely unscreened pan once every so often, just to be sure about what you have.

2. Put the pan fully into the water, shake it a bit, and let it soak for a few seconds. Water is a solvent and it dissolves soil, but sometimes it can take way too long. If you can put the pan all the way in the water, you can spend a few moments with your hands, breaking up any clumps of clay or soil. You can rub the rocks to remove anything clinging to them, and you can pick out any leaves, twigs, or other organic material in there. If you are in deeper, running water, you can hold the pan in one hand and use the other hand to break the material up. Note, however, that there are natural oils on your skin that can build up quickly. If you're using a tub at home, the oils may start to trap very fine gold on the surface of the water. It isn't that the gold is floating, necessarily—it's just trapped in the oils. One way around this is to add to your panning tub a dash of dishwasher rinse aid, such as Finish Jet-Dry. The key to clean glassware in the dishwasher is to break the surface tension, and that's what you want in a panning tub. Just a few drops are all you need; soap will do in a pinch, but it makes for suds.

3. Shake the pan vigorously, at a slight angle toward the riffles so that the heaviest material is sitting at the bottom. You probably can't see the bottom of the pan at this point, so you have to use your imagination to picture what's going on down there. Pour off some dirty water, across

any riffles you have. Don't just dump out the water, because that's using too much energy and you can wash out your heavy material. Let some of the dirty water pour off, and then refresh the pan and see how dirty the water that remains is. Work your fingers in there some more until you can feel that everything is a slurry, or in suspension. This is called liquefaction, and it's crucial because the science of gravity sorting won't work if there are clumps. Pour off some more dirty water, until you feel like you are washing the gravels down to just rock. With too much soil and clay in the water, the difference between the specific gravity of gold and the specific gravity of water narrows from 19:1 to 19:3, and you won't concentrate the heavies at the bottom correctly. So clean water is important.

4. So far, you haven't moved much material out of the pan, but that's OK. These early steps are very important, and doing them right pays off later. Take the pan in both hands and shake it vigorously. Your heavy materials will sink to the bottom, but they'll be spread all over the place down there, across the bottom of the pan. You can swirl and shake and agitate as much as you want, but the key is to get all the material into a liquid state so the heavies settle. Up at the top, where you can see what's going on, you should see the bigger rocks move to a general area, and you can remove them now if they are clean. Orient the pan so you know that the coarsest riffles are opposite you, tilt the pan slightly, and then give it a couple of shakes away from you so that the heaviest material is again at the first riffle. That's where your imagination comes in—you can feel the gravels slide across the bottom of the pan, and you can hear them, too.

   In some parts of the world, placer miners don't use a high-tech, engineered gold pan; they use what's called a batea—basically an upside-down cone—and the heavies always concentrate in the tip. In a metal pan, you can really hear the action, and old-timers talked about the "growl" as nuggets worked across the bottom of the pan. (Many old mining camps in the West were dubbed "Growlersburg" for just that reason; Georgetown, California, was originally named Growlersburg because the miners kept gold-rich quartz specimens in their pockets, and the rocks "growled" as they walked.)

   Let a little material work across the riffles and out the pan. Now you can pick more big rocks out, or scrape them to the opposite side of the

pan away from the riffles. One more trick is what's called the "Blueberry Bounce," named for John Blueberry, an early GPAA member. If you tap the side of the pan, gold will tend to move toward the tap. So if you have material at the bottom of the pan and you don't think it's all near the first riffle, you can tap more of it to that crucial first riffle.

5. Save the swirling action for later, because right now, you don't want to swirl, you want to sluice. You want to use those riffles like a sluice box, and keep sliding material over them and out the pan. Part of the science behind sloughing off lighter material over those riffles is that the water rolling over the riffle causes a little dead zone behind the ridge, and that allows the gold to settle. The other key is to constantly return the pan to level position with a lot of water. This is what the GPAA's Perry Massie calls "re-stratifying" the material. He's a geologist by training, so he tends to use geological terms, but it's pretty easy to understand. Here's a way to think about it: If you were to freeze your pan right now and cut it in half, you would see the different strata, or lines, that you created as you used gravity to sort the contents. The bottom should be thin and black, with the heaviest material, and hopefully with your gold. What you don't want to see is any black sands or gold caught halfway through your pan—they should be at the very bottom.

In Routt, La Plata, and San Miguel Counties, there's a little platinum in those fine black sands, and it's likely to be an important component of your concentrates. Platinum has a specific gravity of 21, while gold weighs in at 19. Other black sands such as palladium, iridium, rhodium, and osmium are also very dense, and in that same range. By contrast, hematite is only 5.6 g/cm^3, and regular rocks such as schist and basalt are closer to 4. Clean, pure water has a specific gravity of 1 g/cm^3. So the science here is that by stratifying your sample, all the succeeding levels toward the top will be consistently lighter and lighter material. But it only works if you have clean water and everything is suspended, not clumped up.

6. Once you have the sample in the pan stratified, slightly tilt the pan away from you and slide the material against that first riffle, with a couple of vigorous shakes and a tap or two. That motion keeps your material at the very bottom of the pan, and you want to keep it as your settling point. One panner I know, Rob Repin of Liberty, Washington, says that if you're panning in a tub, you can now be very aggressive. Rob showed me how he stands the pan up vertically to slide out 95 percent of the unwanted material, because he's confident that he pushes all the good stuff against the riffles. You probably won't have that much confidence, but you should know that science is on his side. With clean water and good stratifying, and a deft slide, you can pack the first riffle and slough out material straight across the riffles and out the pan. Make sure only the top layer of your sample is what moves out—it's the lightest and the least likely to contain any particles of gold that haven't yet sunk to the bottom. You can rock the pan slightly here, too, and watch the bigger rocks roll out.

   On a GPAA outing once, when I was a young greenhorn, I was splashing my pan around and fiercely swirling it like I saw in the movies, when an old-timer had finally seen enough. "Here, can I show you a few things?" he said patiently. He got me to stop swirling the pan and showed me how a gentle rocking motion with the pan tilted away would let the light rocks on top flow out much faster, and much cleaner, than swirling.

7. Stop after two or three sluicing motions and re-stratify the pan, but always keep slamming the good material along the bottom of the pan into that first riffle, and let a little more material out the top as you tip the pan away. You can dip just the end of the pan where the light

material sits and watch it wash away. Use the Blueberry Bounce every so often as well. By now you should see the bottom of your pan, and if you have a good sample, you'll see a line of black sands across the bottom, and maybe creeping up the sides. Sometimes there is so much black sand that you see it very early. But don't start inspecting the pan yet to see if there's any gold in there—be patient, and remove more material. Slough it off, re-stratify, slough some more off, stop, and re-stratify again.

8. Soon, after a series of steps alternately stratifying the pan and pouring a little out across the coarse riffles, you should get down to a few tablespoons of concentrates. Switch to the finer riffles now and repeat your process. Pick out any small rocks that remain, but if they're white quartz or calcite, set them aside to inspect later. You should still shake the contents against the first riffle, and you should see the blonde or tan material rising to the top of the sample, so that they're easy to now rinse across the riffles and out the pan. You can dip the pan carefully into the water now, so that only the part of the riffles with the lighter material is located, and gently wash only that part, leaving the better concentrates alone. Your sample should be getting cleaner and cleaner, but it's going to amaze you how every time you re-stratify, a little more blonde material rises to the top of the sample.

   At this point, if you're using a tub, you don't have to do anything different, because the tub will always catch anything you accidentally lose. But there's an old trick you can use when you get down this far—the safety pan. If you have a metal pan that will sink, you can place it in the creek below you and pan into that from now on. If you don't have a pan that will sink, or the metal pan wants to move, grab a big rock and use that to hold the safety pan underwater. You'll be a little more fearless now, because you know you won't lose anything because you're panning into the lower pan. You can also use that safety pan earlier in the process, and periodically test it to see how you're doing. That's the real trick of gold panning: process, inspect, and test. It's not luck after you get a good process going—it's science.

9. Now pull the pan back level, refill it by letting in a little clean water over the lip, and shake it a couple times from side to side, then slide it against the first coarse riffle like you've been doing. Next comes the fun part: Get the pan level, then lightly tilt the pan toward you so that the contents spread out across the bottom of the pan. This is called a

"fanning" action. Most panners will also cause the materials to slide to the right so that you are creating your own tiny gravel bar, with the heaviest material at the head of that micro-bar. If you're left-handed, you might naturally fan the materials down and to the left. It may take a few tries to get the hang of this, but it's a key part of your process as you get down to the bottom of the pan. This is where having a bigger pan comes in handy, as you'll get a nice, thin spread.

At this point you should see what you have. There may be shotgun pellets rolling around, or even tiny garnets. A hand lens will help identify those. You could see bullet fragments, nails, screws, washers . . . there's lots of junk in some of these waterways. But if you have any decent gold, you'll see it up toward the top, against that riffle. If it spread out too far to the middle of the pan, try using the Blueberry Bounce again and make the gold dance toward your tap. You can now swirl the water carefully across the top, always in the same direction (usually to the right), and wash the lighter material down to the bottom of the pan, a little at a time. This requires practice and patience, too, as it takes a few tries before you learn that the mica flakes in granite country will quickly move around, but the gold stays put. Chances are that the gold you recovered is very fine, but if you have a hand lens, you can see it easily. The bigger pieces will stand out against the black sands and might even wink at you. That's your proverbial "flash in the pan."

10. Congratulations! You have now panned a full sample down to very rich concentrates. You can now make some guesses about where you got your sample from. Are there a lot of black sands? That's good, because this tends to show you are in the right spot. The rule of thumb is that black sands always show up with gold . . . but unfortunately, gold doesn't always show up with black sands, except for possibly at a microscopic level. Are there fishing weights and metal junk? That's good, too, because it shows you have found a trap where heavies are forced to a halt. You're also doing the ecosystem a big favor by cleaning out that garbage, so make sure you don't throw it back in—take it home. Finally, did you see any gold? Then go get another sample! What you want to see is if the next pan comes from a little farther down, and the gold is even better. The deeper you go, the better you should do. Now you're mining.

    You can now slurp up everything in your pan into a snuffer bottle, explained in more detail in appendix A, or you can dump the contents into a separate clean pan to play with later after you've panned down more pay dirt. One way to quickly build up some good concentrates is to stop somewhere during Step 8 and give the material a check, then dump it into a pan. You really don't have to waste a lot of time playing with it, panning it down further, or inspecting it closer. It takes almost as much time to get 90 percent of the material out as it does to get the next 5 percent. And it takes just as much time to get each successive percentage out, down to 1 percent or less, because it requires so much caution and concentration when you're almost done. So you can save a lot of time by checking the pan, saving that, and getting some more material into production. The more material you clean down, the more you'll take home, so the key isn't getting it as clean as possible while you're on the banks of the creek—the key is to keep cleaning pay dirt.

Another key is to check your work. If you used a tub, pour the water off, empty the material into a pan, refill the tub, and see if you missed anything. You can try putting in five little washers or nuts and seeing if you collect them all, but they're really easy to pan after a while. What you really want to do is make sure there are hardly any black sands and no gold in your safety pan or at the bottom of your panning tub. Still, you can't save all the black sand, so don't drive yourself crazy attempting to pan so carefully that none of it escapes.

One more story about panning. I only met George "The Buzzard" Massie, the founder of GPAA, one time, at a GPAA/Lost Dutchman's Mining Association (LDMA) outing. He wasn't down at the diggings, where the heavy equipment was noisily pulling up pay dirt to send through a trommel. He wasn't operating the trommel, or cleaning out mats, or shoveling away buildups. He was up at camp, leaning on their big metal panning trough, going through concentrates with a small crowd around him. He was sipping a light beer and in no hurry, telling a long story about how he and his teenaged boys, Perry and Tom, once worked a dredge on the Stanislaus River, systematically cleaning out a big hole that, by the end of the season, had yielded over 800 ounces of gold. He'd talk a little, then pan a little, continually washing less than a half a gram of black sands out with each push. But the sample got cleaner and cleaner as he talked.

He'd wash a bit, then pull it back and sweep it across the bottom of his big green pan, and everyone who was watching would marvel in delight to see big flakes and small nuggets rolling around. The trommel down at the diggings was moving at least 5 yards of material per hour, and each time the crew pulled the mats and cleaned up, The Buzzard would get another bucket of concentrates—he called them "cons"—to work on. You got the feeling he could use a pie tin or a frying pan and get the same results, and at one time, he probably had. Productivity consultants say it takes 10,000 hours of practice to become an expert, and you could just tell he'd surpassed that benchmark a long, long time ago.

Every once in a while, a tiny flake would go over the lip of the pan, but he said it just made the big tub all that richer when they cleaned it up some day. What he really showed us was patience, living in the moment, and enjoying the simple act of cleaning your sample and telling a good story. He died of a heart attack a few months later, but his legacy was already established. He had built a nationwide club with claims in just about every major district, at a reasonable yearly price, with dozens of state chapters. His GPAA sponsored a full season of mining shows across several states, founded a television network devoted to outdoor activities, and provided a loud voice standing up for the rights of people to enjoy panning on public lands. He was a great man, and to me, he seemed happiest when he had a pan in his hands.

MAP LEGEND

70 Interstate
550 US Highway
9 State Highway
County/Forest/Local Road
Unpaved Road
Trail
Railroad
Forest Boundary
Park
Small Park
1 Trailhead
Mine
Museum
Point of Interest
Camping
Primitive Camp
Viewpoint
P Parking
Picnic Area
Gate
Boat Ramp
Ski Area
Ranger Station
Airport
Cafe
Waterfall
Spring
Mountain/Peak

Part I: Southwest Colorado

1. Colorado River

Land type: Riverbank
County: Mesa
Elevation: 4,464 feet
GPS: 39.17548, -108.80747
Best season: Late fall for low water; avoid desert heat in July and Aug
Land manager: Horsethief Canyon State Wildlife Area
Material: Fine gold
Tools: Pan only; hand tools OK
Vehicle: Any. The gravel road is a bit rough, but even slow-moving sedans should be OK.
Special attractions: Colorado National Monument; Museums of Western Colorado in Fruita
Accommodations: No camping at site but available nearby at Highline Lake State Park. RV parking and motels in Fruita and Grand Junction.
Finding the site: From Fruita, go 5 miles west on I-70 to the Loma exit (exit 15) for CO 139. Drive to the south side of the interchange and turn left toward 13 Road. Drive east and then south on 13 Road about 0.4 mile to the parking area. The coordinates were taken at this main parking spot.

Prospecting

Most of the gold you'll find here in the Colorado River probably comes from the Gunnison River, so the advantage of this location is that it's below the mouth of the Gunnison, which joins the Colorado in Grand Junction. There are some minor districts in the Colorado drainage at Rifle and beyond, but most studies don't show serious interest in the river until it reaches the Grand Canyon.

Still, this locale offers easy access, even if it isn't on a good bend. The gold is extremely fine, and there isn't a lot of area to work with, but you should be able to pan a few colors. If the boat launch isn't swarming with rafters, you can try the notch in the bank just downstream from the launch area. When the water is low, you can see where the river tried to create a new path, and there is some decent cobble there. Or, upstream about 100 yards, you'll find a little bend in the river and some good rocky areas. You can reach another area

When the water is low, you should be able to pull some colors from the Colorado River. Walk about 100 yards upriver; do not disturb the banks or dig up the beach at the boat launch.

from a path heading down from where you park. The mouth of the Kiefer Extension of the Grand Valley Canal isn't a good place to dig, as the banks are unstable and ready to collapse. Also, the quality of the water flowing in from this canal is probably not high.

Look for garnet and zircon in your pan once you get a few heavies to check. It's always a good idea to examine your black sands with a loupe or hand lens to get a better look at what you recover.

The beach area can get busy, and parking at the river is discouraged. There are additional access points on the Colorado in the McInnis Canyons National Conservation Area, but they are best accessed by mountain bike, such as via the Kokopelli Trail. In Grand Junction you can also locate a few access

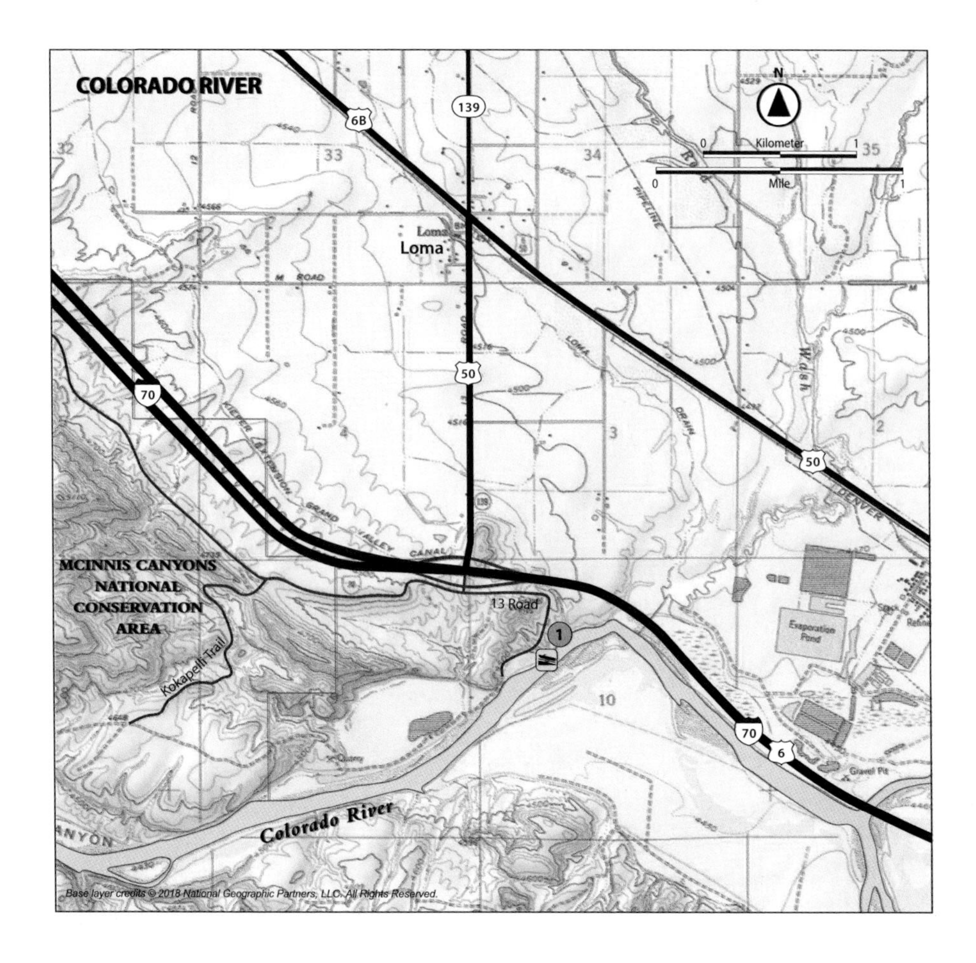

spots, such as off Railroad Avenue at 39.10551, -108.65076 or off Redlands Parkway at 39.08902, -108.61838. Be prepared to hike to get to a good river bend, and look for natural traps among boulders or cobbles.

2. Rattlesnake Gulch

Land type: Riverbank
County: Delta
Elevation: 4,852 feet
GPS: 38.74153, -108.22247
Best season: After summer heat for low water
Land manager: Dominguez-Escalante National Conservation Area
Material: Fine gold
Tools: Pan, hand tools
Vehicle: 4WD suggested; final approach is rough.
Special attractions: Black Canyon of the Gunnison National Park; Dominguez Canyon
Accommodations: Primitive camping at site; developed campground at Sweitzer Lake State Park in Delta. Motels and RV parking in Delta.
Finding the site: From Grand Junction, drive 28.8 miles east on US 50, then turn right onto Rattlesnake Gulch Road/730 Road. If coming from Delta, drive 10.4 miles west on US 50. (Your GPS may suggest the eastern connection to 730 Road, but it looks even rougher than the western side, and I didn't use it.) Follow 730 Road as it winds south, then drive east along the Gunnison River and the railroad tracks. You'll reach a big turnaround area, with some decent shady areas to camp at, but this isn't the end of the road. One option parallels the river and the tracks, and gets rough in places. If you continue on 730 Road, it also gets rough. They eventually meet at the only decent railroad crossing. Cross the tracks and continue to the river; the road forks, but both routes take you to the Rattlesnake Gulch BLM sign. The GPS coordinates were taken close to the sign.

Prospecting

Rattlesnake Gulch used to be a popular spot for recreational gold prospecting, with dredging, sluicing, high-banking, and lots of digging into the bank. That has all changed, and all the BLM now tolerates is gold panning, as long as it's "casual use" only. That means no holes in the bank, no dredges or high-bankers, etc. Stick to gold panning and hand tools, collect material from the water line only, and concentrate on working around the big rocks in the water. Those giant holes you see where energetic prospectors of yesteryear

Once you spot these old diggings, you'll know you've reached the open panning area at Rattlesnake Gulch. Note that digging big holes like this is no longer allowed.

dug into the big piles of rock and cobble are just relics of the past, and not to be duplicated.

This area does still yield mostly fine gold, but it can be extremely hot here in the summer when the water finally goes down. You can try working closer to the trees, to the west, but the hydraulics aren't great there, and you're just going to see flood gold. What you want to try to do is get a hole going in the river bottom, at or below the water line.

There were historic placers at Hotchkiss, on the North Fork of the Gunnison, but that's now a sand and gravel operation. The gold at Rattlesnake Gulch probably originates far to the east, in the Gunnison District, Beaver District, and Willow Creek District, and from Cebolla Creek. Farther east, Quartz Creek and Tomichi Creek flow into the Gunnison drainage near Parlin, but the dam at Blue Mesa restricts any further replenishment from those sources.

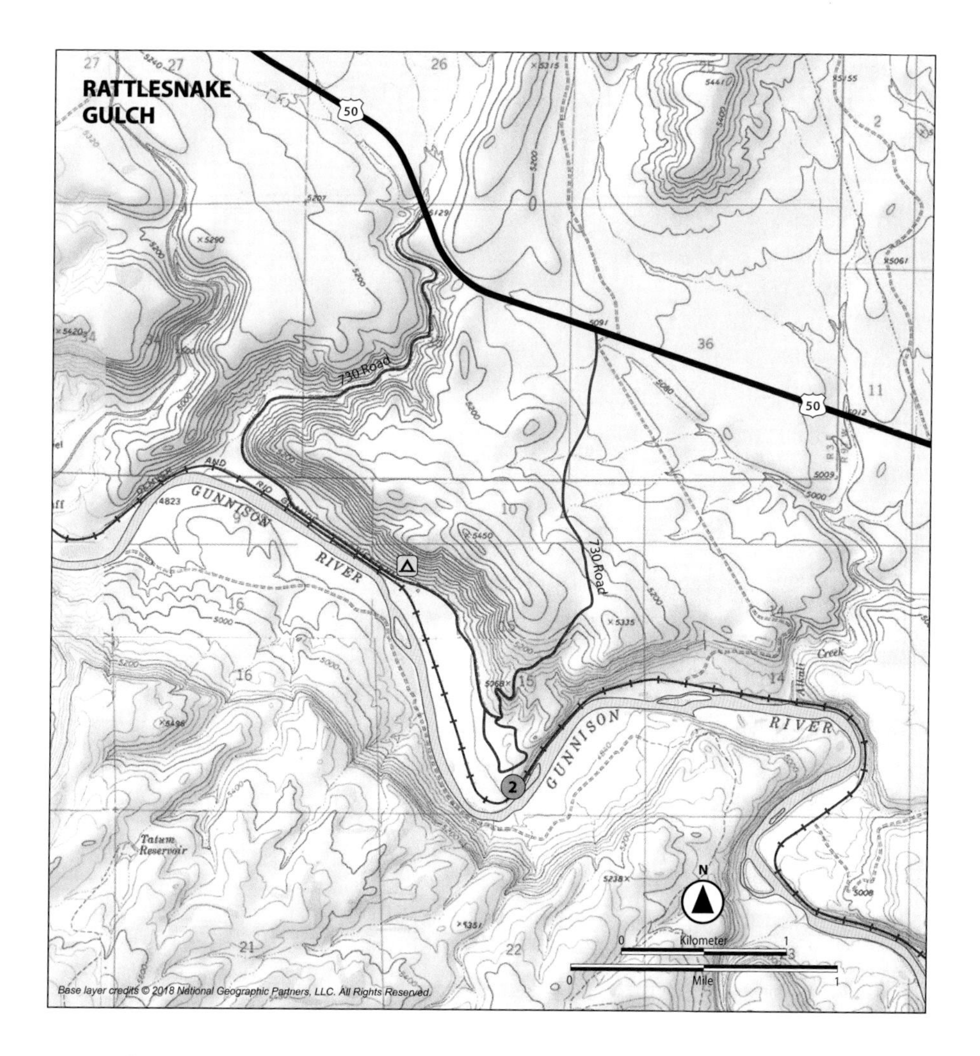

The nearby Black Canyon of the Gunnison National Park is a fascinating view of ancient rock formations. Hansen (1965) describes formations dating to the Precambrian, including gneiss, quartz-mica schist, amphibolite, and quartzite. Immense sections of sedimentary rocks cap the older horizons, and they're easy to learn here so you can recognize them later—including the world-famous Morrison Formation, host to so many dinosaur fossils, and the Dakota Sandstone, also full of fossils.

3. Lower Dolores River

Land type: Riverbank, dry gulches
County: Mesa
Elevation: 4,477 feet at lowest point
GPS: A - Rapids: 38.71638, -109.01308
B - Shady camp: 38.72171, -109.02006
C - Wash: 38.72536, -109.02096
Best season: Late summer and early fall for low water
Land manager: BLM
Material: Fine gold
Tools: Pan, sluice
Vehicle: 4WD recommended; rough road in places
Special attractions: Palisade Butte
Accommodations: Primitive camping; limited services in Gateway
Finding the site: Cross the Dolores River to the south side at Gateway, and look for a right turn onto 4 1/10 Road. Site A is an access point at the rapids about 3.7 miles away, close to Utah, with a few farms and fences blocking good access to the water before this spot. This locale is a diversion dam, so don't mess with the big rocks. There are some good spots where the gulch enters in, below the dam, and against the cliff, plus downstream. Site B is about 0.5 mile closer to the Utah border, and is a shady primitive camping area. Site C is a dry wash, about 0.4 mile east of Site B.

Prospecting

The lower Dolores is sparsely populated, but there are ranches along the route to the state line. You don't need to cross any fences or trespass to reach these access points, so be a good citizen and stay on open BLM land. There is a Gold Prospectors Association of America (GPAA) claim called "Gold Creek" at the Utah state line. The claim starts at about 38.74289, -109.03932 and lies to the west, covering the final three rapids. I didn't see any claim posts or signs elsewhere, but always be on the lookout for new signage.

When the water is lowest, you should have luck around some of the lower bends in the river, especially between Site B and Site C. At Site A you can park easily and explore the beach area below the rapids, and if the water level

The lower Dolores River carries loads of black sands and very fine gold, mostly picked up from the San Miguel River drainage.

has dropped dramatically, you should see black sands everywhere. At Site B there is primitive camping, and during low water you might be able to check the eastern tip of the islands forming there. At Site C the river makes a good bend at a dramatic pinch point, and there is also evidence of primitive camping here.

Sometimes the gulches that enter from the south act as riffles in a sluice box when the water is high, and you might want to try a metal detector in them. Just be aware that there is a lot of trash to dodge.

The gold here is very fine, having traveled from as far away as Telluride and Rico. You should find loads of black sands, masking the tiny colors. To get larger pieces of gold, you'll need to get a good hole going around some large boulders. Bedrock is not easy to find. The Dolores makes a big bend through southwest Colorado, but due to the dam at Dry Canyon, forming McPhee

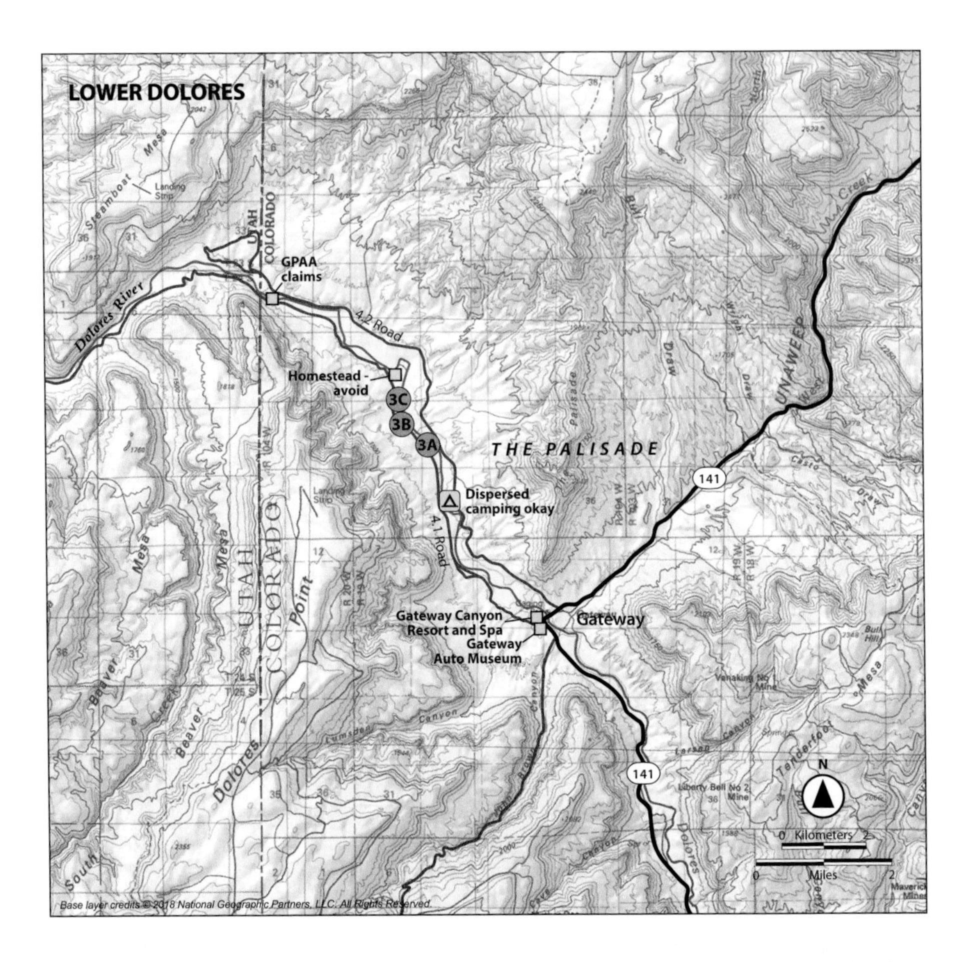

Reservoir, it won't be replenished like a free-flowing river. Fortunately, the San Miguel River still flushes gold down from the Telluride area, and it empties into the Dolores below Uravan.

4. Uravan

Land type: River canyon, riverbank
County: Montrose
Elevation: 4,744 feet at lowest point
GPS: 38.37987, -108.80281
Best season: Summer and fall for lowest water
Land manager: BLM; avoid private ranches
Material: Fine gold
Tools: Pan, sluice, dredge, high-banker
Vehicle: 4WD suggested for Y11 Road; steep and bumpy in stretches
Special attractions: Hanging Flume; Hieroglyphic Canyon
Accommodations: Primitive camping along river; full services in Naturita and Gateway.
Finding the site: Head east from Gateway on CO 141 for 35.5 miles. Be sure to stop at the interpretive signs erected to describe the flume and uranium mining. If coming from Naturita, drive 15.3 miles west. Cross the bridge at what's left of Uravan and take a right onto Y11 Road. Drive 5.1 miles to a pullout where you can park by the river; the coordinates were recorded at the parking spot. This road isn't too bad, and it goes through to Bedrock and makes a nice shortcut to the La Sal District.

Prospecting

The Hanging Flume near Uravan was an engineering marvel built by Nathaniel P. Turner from 1889 to 1891. The goal was to deliver enough water from the San Miguel River to power a hydraulic mine about 4 miles downriver from where the Dolores and San Miguel join. Bolted into canyon walls, the flume did finally reach the Bancroft Claim, but the Montrose Placer Mining Company was not profitable and soon went out of business. Builders tried to extend the flume to the Vixen claims at 38.43694, -108.83844, but that effort also failed. You can read all about it at www.hangingflume.org.

Today you can see the old ruins of the flume from a pullout off CO 141 at 38.39852, -108.80934, but the best view is from the river, via River Road (Y 11) at 38.38476, -108.79235. In 2012 a crew reconstructed a 48-foot section, which you can view from here. Thanks to the JM Kaplan Fund, the Hendricks Family Foundation, and the Colorado State Historical Society, among others, along with the support of the BLM Uncompahgre Field Office

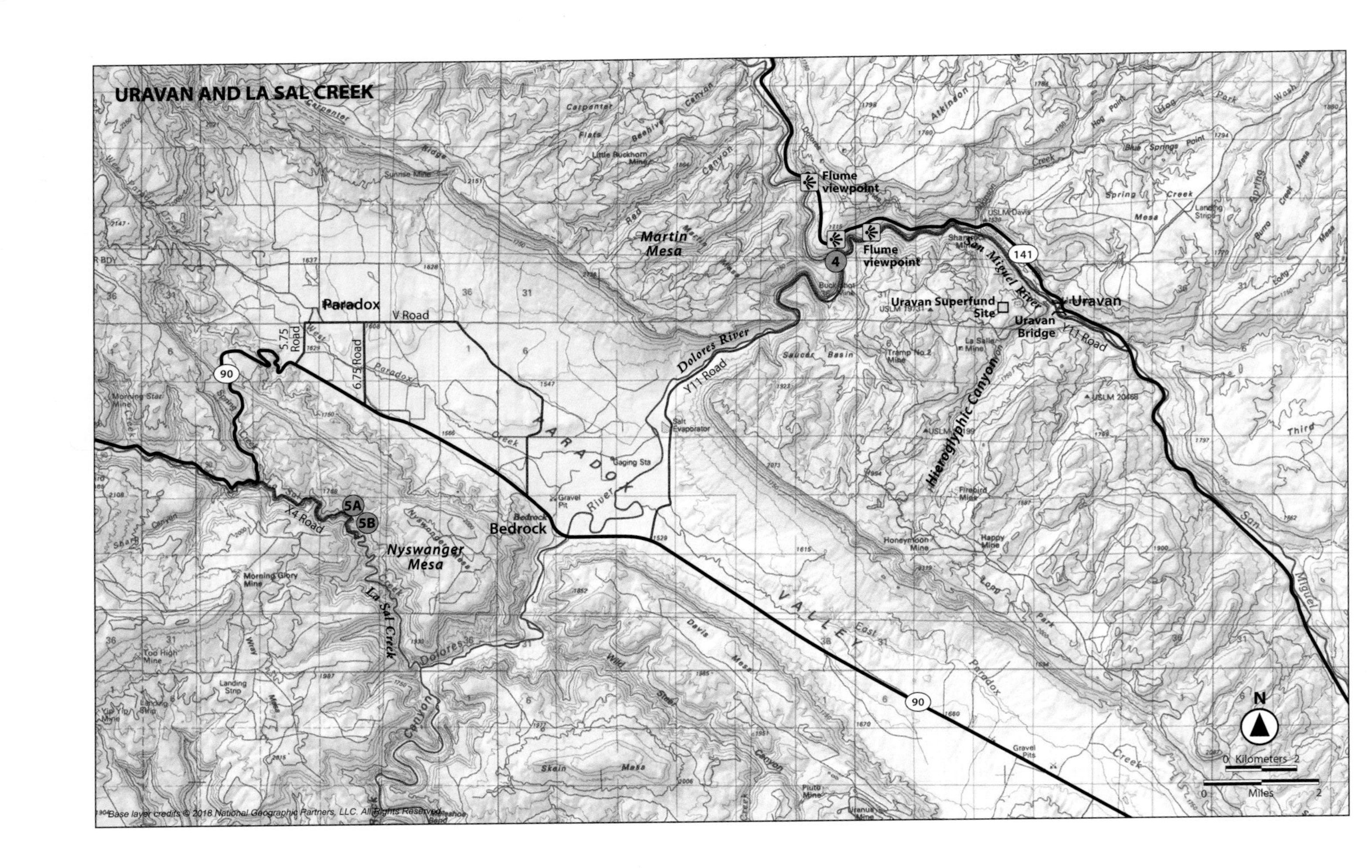

Base layer credits © 2018 National Geographic Partners, LLC. All Rights Reserved.

This partially restored section of the Hanging Flume shows there was enough pay here to fund some amazing engineering feats.

and Western Colorado Interpretive Association, the flume has been nominated to the World Monuments Fund Watch list. Obviously, this means you should not do anything to damage the flume, and any historic artifacts you find nearby should be turned over to the BLM immediately. Record the GPS coordinates if you do find something interesting.

The area of the Dolores River that the flume sought to exploit is on a mixture of private land, current mining claims, and some open spots. Very few claim markers are visible, but the current GPAA Claims Guide lists five claims covering 680 acres spanning some of the area from Blue Mesa to Salt Creek. They are known as Reece's Honor and CSC1-4.

I've never had much luck panning the Dolores River at its confluence with the San Miguel. Mostly I've recovered a lot of blonde sands and very little black sand, and only tiny pinpricks of gold. The San Miguel, on the other hand, offers loads of black sands and much better color. The San Miguel River not only pulls gold from the drainage at Telluride, but also picks up mineralization from along the canyon. Most of the mines out here are for uranium or vanadium, but gold, silver, copper, and other metals are present in the canyon walls.

There is a good pullout at the mouth, and some big rocks to scoop up dirt around. The San Miguel forms some interesting bends just before its mouth that are worth checking, and the old flume is hanging from the canyon wall above you. Be sure to check out the plaque.

You might also go a bit upstream on the Dolores and see how different the gold is from the two drainages. If you find a good pay layer, you should notice that the river has finer gold, as it has traveled farther.

5. La Sal Creek

Land type: Seasonal creek, dry gulches
County: Montrose
Elevation: 5,410 feet
GPS: A - Cliff Dweller Mine: 38.31585, -108.95244
B - Cashin Mine: 38.31071, -108.95008
Best season: Spring offers best chance to have water for a sluice.
Land manager: BLM
Material: Fine gold
Tools: Pan
Vehicle: 4WD required
Special attractions: Paradox Valley
Accommodations: Primitive camping here with little shade. Closest facilities are at Nucla and Naturita, with an RV park in Naturita.
Finding the site: From Naturita, take CO 141 west for 2.1 miles, then turn left onto CO 90. Drive 29.5 miles and look for the left turn onto X4 Road. Drive 3.2 miles on this rough road to Site A. Site B is at the end of the road, about 0.6 mile farther.

Prospecting

The gold here is very fine, and is a by-product of more concerted efforts to find copper in the cliffs. Both the Cliff Dweller Mine and the Cashin Mine produced copper, with associated gold, silver, and zinc, and there was enough mineralization to build a pretty good structure and create a serious mine. Use extreme caution around the structures, and stay out of the underground workings. We crossed the creek on an old, rotting bridge, then found an easier ford across the creek closer to the dead-end in the road.

Keep an eye out for any obviously mineralized rock, which will appear blackened as the sulfides oxidize. You should be able to easily find some greenish malachite; the blue azurite is a bit harder to find on the dumps, but after a while, you should spot some. During the hot months, you'll be rewarded with a blast of cold air howling out of the main adit.

There are multiple natural traps along the creek, and once you are close to Site A, you should start sampling. Look for any big rocks you can dig around, and use the natural traps as the creek bends. The creek runs slow and sluggish

The crumbling Cashin copper mine was the heart of the La Sal District.

by late summer, and your best bet is to follow the water past the mine and look for bedrock cracks to clean out. The gold is spotty and fine, but a little effort should be rewarded.

6. Piñon Bridge

Land type: Riverbank
County: Montrose
Elevation: 5,864 at lowest point
GPS: A - Piñon Bridge: 38.26327, -108.40004
B - Cottonwood Ledges Campground: 38.24689, -108.38197
C - Rock House Ledges Campground: 38.23483, -108.36914
Best season: Summer for lower water, but gets hot
Land manager: BLM. Recreational Prospecting permit required (call ahead to the Uncompahgre BLM Field Office in Montrose at 970-240-5300). You'll need the BLM map.
Material: Fine gold, occasional small flakes
Tools: Pan, sluice; nonmotorized recreational panning
Vehicle: Any, but sturdy vehicle suggested
Special attractions: Telluride
Accommodations: Multiple semi-developed campgrounds throughout the area. High Country RV Park in Naturita. More developed campgrounds at Telluride, plus RV parking, motels, and lodges.
Finding the site: From Naturita, drive east on CO 141 3.2 miles and turn left onto CO 90. Drive on this main road for 7.2 miles. To reach Site A, travel east. After about 0.8 mile, you'll reach Piñon Bridge. Turn hard right just past the bridge to leave CO 90 and start on BB36 Road. After 0.3 mile the public area begins, going upstream. This is Site A, and it is open for about 0.1 mile. Site B is about 2 miles upstream, and there is primitive camping here. Site C has more camping and access, about 0.6 mile farther upstream. The public area ends at 38.22035, -108.35166.

Prospecting

Historically there were several placer mines east of Naturita along the San Miguel River: The Amalgamator Flat property was near the CO 97 bridge, and the Philadelphia, Annie, Rose, and Mary Ann Placers were near where EE30 Road swings off of CO 90. Unfortunately, most of the land here is private and access is iffy. So your best bet, unless you know someone, is to stay on CO 90 and drive to the Oro District, near Piñon Bridge. There were several placer operations here back in the day, including the Lauer, Rock Ford, O'Brien, and Miracle Placers, comprising the Oro District. The source

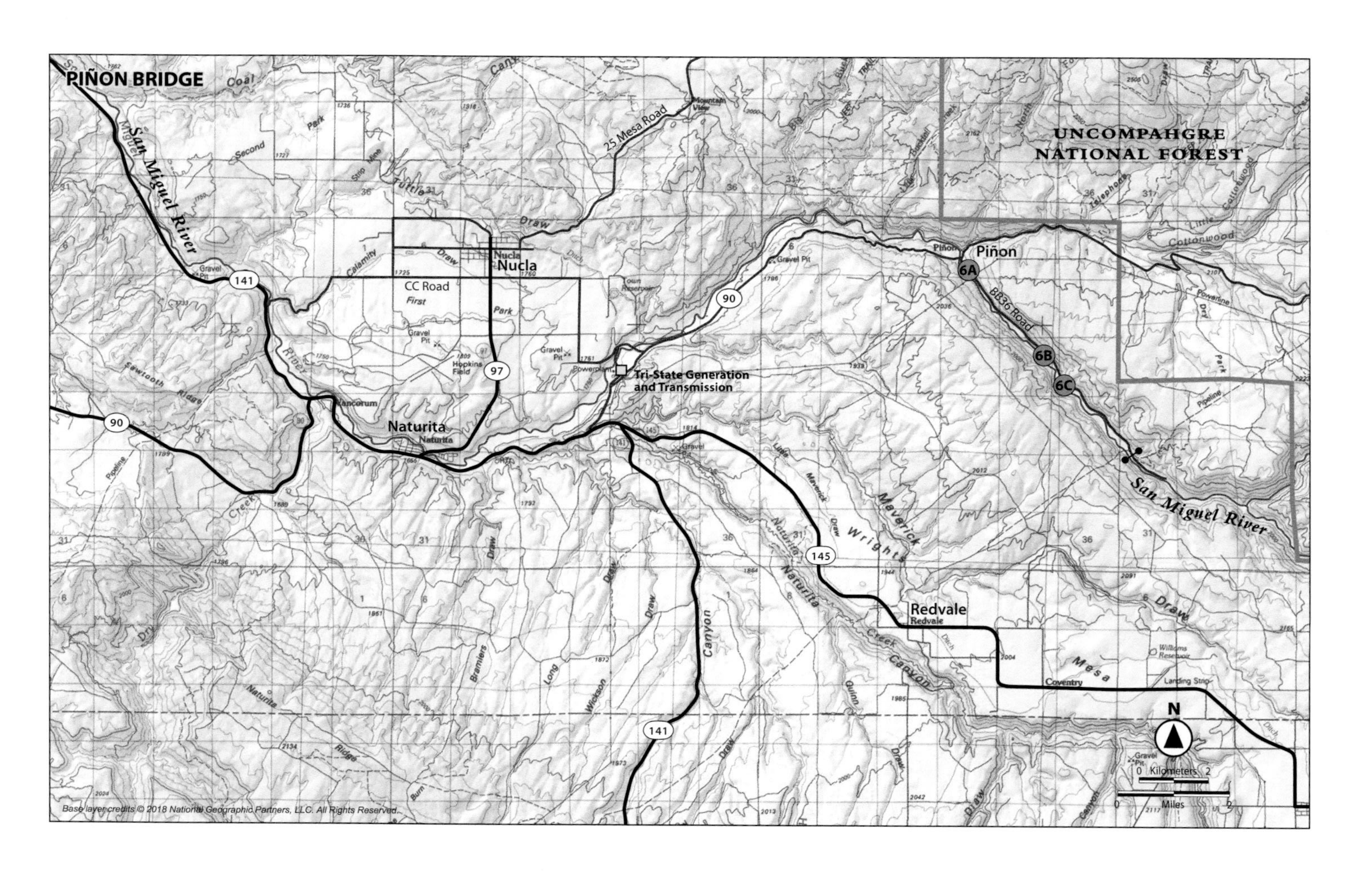

Base layer credits © 2018 National Geographic Partners, LLC. All Rights Reserved.

This area is crowded with more rafters than prospectors during much of the summer rafting season, but it is open for casual, recreational prospecting. Steer clear of claims—get the map and permit before you venture out!

for the gold is located either in the Telluride Mining District, about 50 miles upstream, or in some of the mines that dot the canyon walls near Placerville. So the gold is very fine, but flakes aren't impossible to dig up if you find a good pay streak among the larger boulders.

Note that you need to contact the BLM and get a signed permission form, which you can easily do via email after you call them. That goes for Norwood Hill as well.

Site A is near what was once the Miracle Placer Mine, but it's along a very straight stretch of the river. There are some decent boulders to play around here, once you blaze a trail through the trees. The camping area at Site B offers good access to the water, but shade is limited, and it will obviously be hit harder. An abandoned river course ends just upstream of the outhouse, and the upper, beginning section of this old channel is interesting. The river is open upstream from the camping area for about 0.1 mile. Similarly, the river is open upstream from the camping area at Site C to almost the end of the road, where the ranching begins. There are some good bends in the river here to explore. This road does not continue through, so you'll have to backtrack to Naturita to reach the Norwood Hill site.

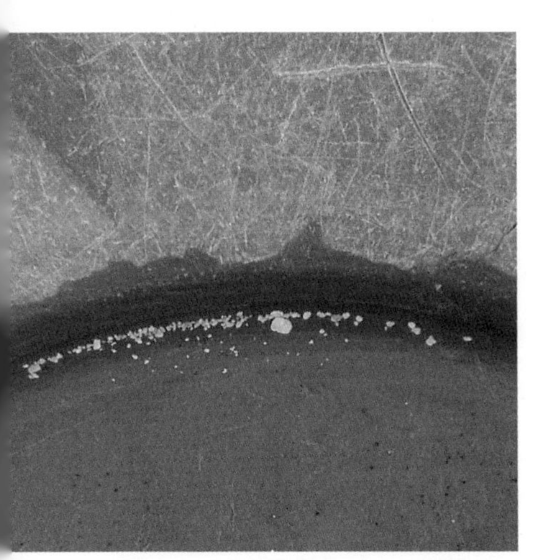

San Miguel gold from the Piñon Bridge site after about three hours digging a hole.

7. Norwood Hill

Land type: Riverbank
Counties: Montrose and San Miguel
Elevation: 6,621 feet at bridge
GPS: A - West: 38.12574, -108.20634
B - East: 38.11558, -108.19541
Best season: Spring for more water, less heat; later in fall for cooler temps
Land manager: BLM
Material: Fine gold
Tools: Pan, hand tools
Vehicle: Any
Special attractions: Telluride
Accommodations: Primitive camping not allowed here; try USFS land along Sanborn Park Road above Site A. Full facilities, including campground, RV park, and hotels, in Telluride.
Finding the site: From Cedar Street in Norwood, drive east on CO 145 for 5 miles. On your right is a large parking area and a gravel road that leads up to the public parking on the south side of the river. Or you can proceed another 0.1 mile across the bridge farther east, to the parking area on the right. This is Site A; the river is open above here. To reach Site B, drive another 0.5 mile to a decent parking area on the right. Note this spot, then drive another 0.4 mile upstream to the coordinates, where a major dry gulch runs in from the northeast and under the highway. This is Site B. There have been claim markers above here in the past.

Prospecting

While the land ownership is easier to figure out here compared to Piñon Bridge, this stretch of the San Miguel River flows fast and relatively straight, and there are few easy pull-offs. Site A at the bridge offers a good parking spot, but steer clear of the bridge itself and stay upstream. There are a couple more parking spots about midway through the recreation area and a small one at the end. Other than that, parking is limited. Camping is probably OK, but there are very few places to pitch a tent unless you come in from the hiking trail on the west end. Pull well off the pavement if you stop anywhere that isn't a proper spot.

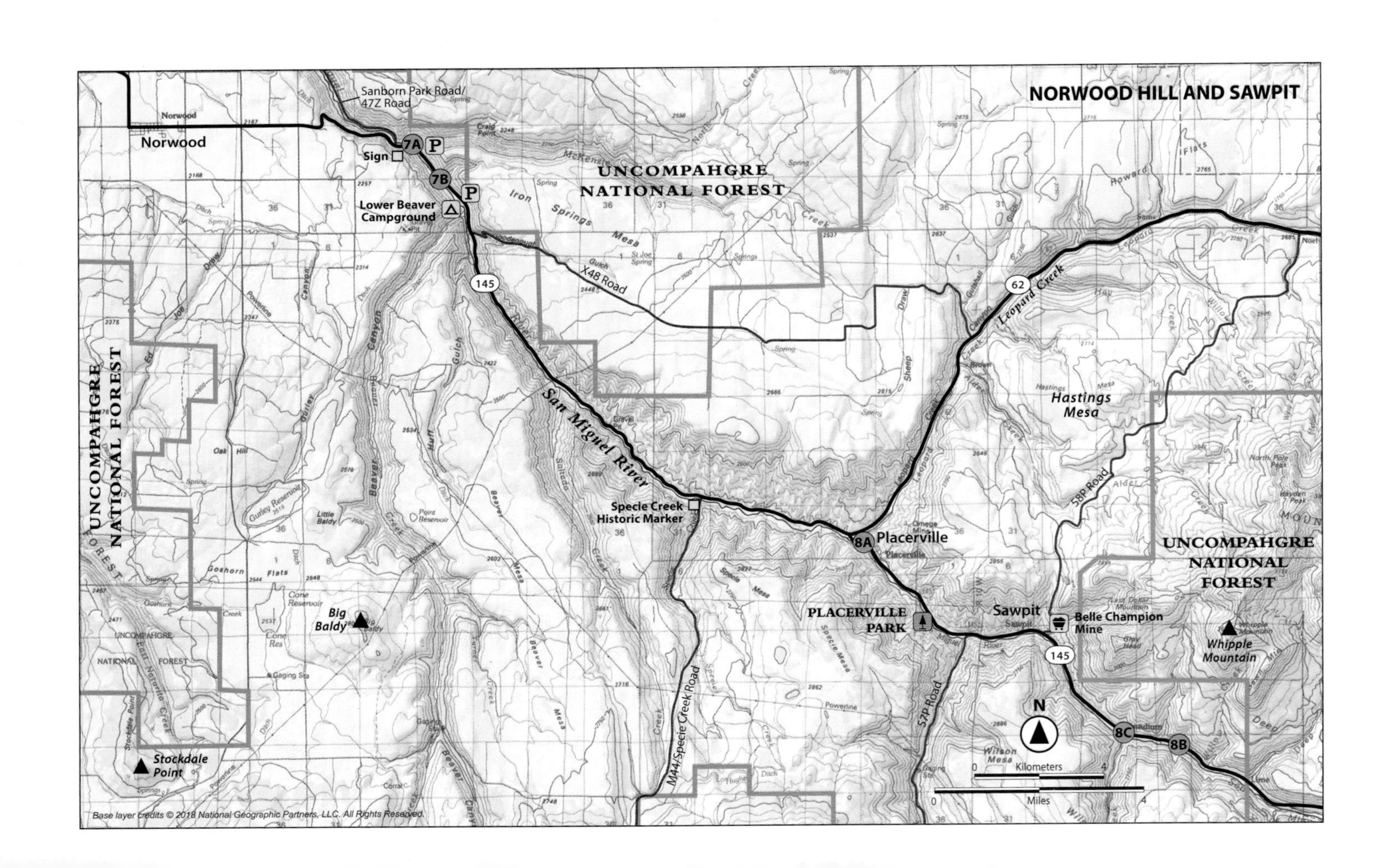

Base layer credits © 2018 National Geographic Partners, LLC. All Rights Reserved.

There are a few bends in the river to inspect at Norwood Hill, so be patient and thorough.

The gold here is deep, and you'll have to locate some big rock traps or visit during very low water, late in the year. There are some promising bends in the river, and more rocks pop out when it's running low.

Another nearby spot is at 38.10779, -108.18696 near Lower Beaver Campground. Be sure to look around for claim markers here, but the campground is typically open for panning, as are most along the river here. Another nearby area that may not be under claim when you get there is at Clay Creek, at 38.15985, -108.24191. Drive east 3.7 miles from Naturita on CO 131 to the junction with CO 145 and turn left onto CO 145. Drive 20.7 miles to 47Z/Sanborn Park Road and turn left. After 3.3 miles a corner of USFS land has access to the water but is open for claims, so check for signage. Sanborn Park Road veers right, up the hill, headed for Sorenson Reservoir and other seasonal reservoirs just past the big bend in the road. Seasonal Clay Creek enters the river here.

8. Sawpit

Land type: Riverbank
County: San Miguel
Elevation: 7,303 feet at Site A; 7,757 feet at Site C
GPS: A - Placerville: 38.01826, -108.05577
B - Public access: 37.96376, -107.95296
C - 60M Road: 37.96707, -107.97064
Best season: Spring for runoff and cooler temperatures; late fall for lower water
Land manager: San Miguel County
Material: Fine gold
Tools: Pan, hand tools
Vehicle: Any; very few side roads
Special attractions: Specie Creek hydraulic mining heritage site (38.03038, -108.11219)
Accommodations: Multiple developed campgrounds along the river; full services in Telluride.
Finding the site: To reach Site A from Naturita, drive east on CO 141 about 3.6 miles, then take CO 145 for 32.5 miles. Turn right to stay on CO 145S, and drive about 0.4 mile to the rafting access beach. If driving from Telluride, head west from the intersection of Fir and Colorado about 0.5 mile, keep going west on CO 145 for 2.9 miles to the next traffic circle, then drive 12.4 miles to the rafting access beach. To reach Site B, backtrack on CO 145 toward Telluride for 7.5 miles and look for a good pullout from the highway. Park carefully as far from traffic as possible, and scamper down the hill to the water. Alternatively, if you don't mind a good hike, you can drive 1 mile east from Site B to the turn for 60M Road. Turn south and locate a large parking area with a big sign on the gated road to the west. Hike a mile west and make your way to one of the modest bends in the north bank of the river.

Prospecting

The Sawpit-Placerville area saw considerable placer mining from 1875 to 1882 (Parker 2009, 76). Much of the area is now in private hands, and there are a few active claims in the area to dodge as well. The recovery was never as good as hoped for here; you'd expect that being so close to the major mines

This new public panning area requires either a 1-mile hike or a scamper down the hillside from the parking area coordinates for Site B.

at Telluride, the gravels of the San Miguel would be fairly rich. But Parker indicates that the terraces and benches from 80 to 385 feet are more interesting than the current river gravels. The bedrock here is sandstone, but it's not all soft, and the terraces sit on harder sandstone with considerable gravel and cobble piled up. You'd have to fill buckets and bring them down to the water in order to exploit them, which is a lot of work.

The rafting access at Placerville sits on the site of the old Rosemary Placer, which wasn't under claim in 2017. Keep an eye out to see if that has changed, but you should be OK for a few test pans. Try to get a decent hole going, and always be sure to fill your hole back in. Working your way upstream, you should see some very small gravel bars forming and an abandoned channel as well.

Sites B and C both access the same public panning area recently opened by the BLM. The signs indicate it is temporary, so if new signs pop up, you'll know if the status has changed. It might be a good idea to check Site C first to verify that the sign is in place, then take your chances finding good parking at Site B. The hike in is very pleasant, being just slightly uphill, but it's a long walk if you are bringing in lunch, panning tools, etc.

Placerville County Park at 37.99883, -108.03475 also offers good access to the river from athletic fields; it's a short hike, and there are plenty of large rocks to move around when the water is low. It just feels a little weird packing a pan over to the river while everyone else is playing soccer or softball. The Specie Creek picnic area at 38.03038, -108.11219 has some interesting historic signs for the old hydraulic mine here. You can still see the scars on the hill to the southwest. One of the information signs provides a quick primer on panning for gold, and you can give the river a try at the lower access site.

Nearby Wheeler Bar at Sawpit is private, so there's no access there. On the east edge of Sawpit, 58P Road takes off north into the mountains and provides access to the Lizzie G Mine, mostly obliterated by the road construction, and the Belle Champion Mine. You can access the Belle Champion Mine via a dirt road from the hairpin curve. The coordinates are roughly 37.99861, -107.99509, and the tailings are red, gray, and buff.

After the gold ran out, Placerville was noted for vanadium deposits; Chenowith (1980, p. 217) reported that in 1919 about 30 percent of the world's vanadium came from at least five active mines in the area. About 3.7 million pounds of vanadium originated at Placerville by 1940, along with associated uranium. Be sure to avoid any adits and stay out.

There are more access points between Sawpit and Telluride, but as always, be on the lookout for fences, private property, and claim markers. Don't be a nuisance—that's why public areas were created, to give you a safe spot to explore.

9. Telluride

Land type: Creeks, mountains, tailings
County: San Miguel
Elevation: 8,803 feet in downtown Telluride; 11,404 feet at Tomboy Mine
GPS: A - South Fork: 37.94227, -107.89861
B - Tomboy Mine: 37.93874, -107.75899
C - Gold Creek: 37.88504, -107.85485
D - Museum: 37.93963, -107.81084
Best season: Fourth of July through early summer
Land manager: San Juan National Forest
Material: Fine gold, rare small flakes. Look for interesting mineral specimens on the Tomboy tailings.
Tools: Pan, sluice
Vehicle: Any for Sites A, C, and D. Site B requires 4WD—the drive is up a steep one-lane mountain road with barely enough turnouts; remember that uphill drivers have right-of-way.
Special attractions: Downtown Telluride offers shopping, dining, brewpubs, and more.
Accommodations: Sunshine Campground (37.88933, -107.88848) open seasonally. Primitive camping available in much of this area, if you stay well away from private property. Hotels, resorts, RV parking, and vacation rentals in Telluride.
Finding the site: Telluride is about 16 miles east of Placerville. To reach Site A, drive from the middle of Telluride at Fir Street and Colorado Avenue for 6 miles west to CR 673L; turn right and follow 673L for 1.2 miles. Turn right and either park here, if the gate is closed, or wander through the sparse forest on the main track. The coordinates are from a large parking area not far from the water. The South Fork of the San Miguel is on your left, and it joins the main stem just ahead of you. To reach Site B, you'll need a good 4WD and nerves of steel. From downtown Telluride at Fir and Colorado, drive north on Fir for 2 blocks to Columbia, turn left, go 2 blocks west on Columbia to Oak, and drive north on Oak until it turns sharply right and becomes Tomboy Road. Go steeply up the mountain 4.4 miles. Uphill drivers have the right-of-way on this narrow road, as it's far harder to back up safely down a hill. Always remember where the last turnout was, and try to get into a convoy so that you have strength in numbers. There are extensive mine ruins along this ridge, so

Driving up to the Tomboy Mine ruins is a popular summer activity out of Telluride. There are multiple places to rent a Jeep or ATV for the drive.

drive around once you're up at 11,000 feet. To reach the old Warren Placers on a tributary of Gold Creek at Site C, start from downtown Telluride at Fir and Colorado and drive west on CO 145 for 3.4 miles. Zip through the traffic circle to end up with a left turn, driving on CO 145S. Go 5.3 miles and turn left onto Alta Lakes Road (this road is closed during much of the year). Go 3.6 miles to the beginning of the ruins and look for the creek on the right side of the road. New signs up there advise you to stay off the old tailings piles, but there are lots of picturesque ruins to capture with your camera. Finally, Site D is the Telluride Historical Museum, which you can reach by driving north on Fir Street to Gregory Avenue.

Prospecting

Telluride is one of Colorado's gems, offering a rich mining history that you can only truly explore during the summer season. The town is famous for its ski runs in the winter months, but mining is what put Telluride on the map. In 1875 silver ore from the Smuggler Vein started producing riches; in 1882 a shipment of four tons of high-grade ore yielded 800 ounces of silver and 18 ounces of gold per ton. The town quickly grew, changed its name from Columbia to Telluride, and even attracted the attention of Butch Cassidy, who robbed the San Miguel Bank in 1889 and made off with almost $25,000 in his first big robbery.

The Tomboy, Pandora, Smuggler-Union, Nellie, and Sheridan Mines were the most important producers, and you can still pick up interesting specimens at many of the ruins. Although the town is named for an interesting gold ore,

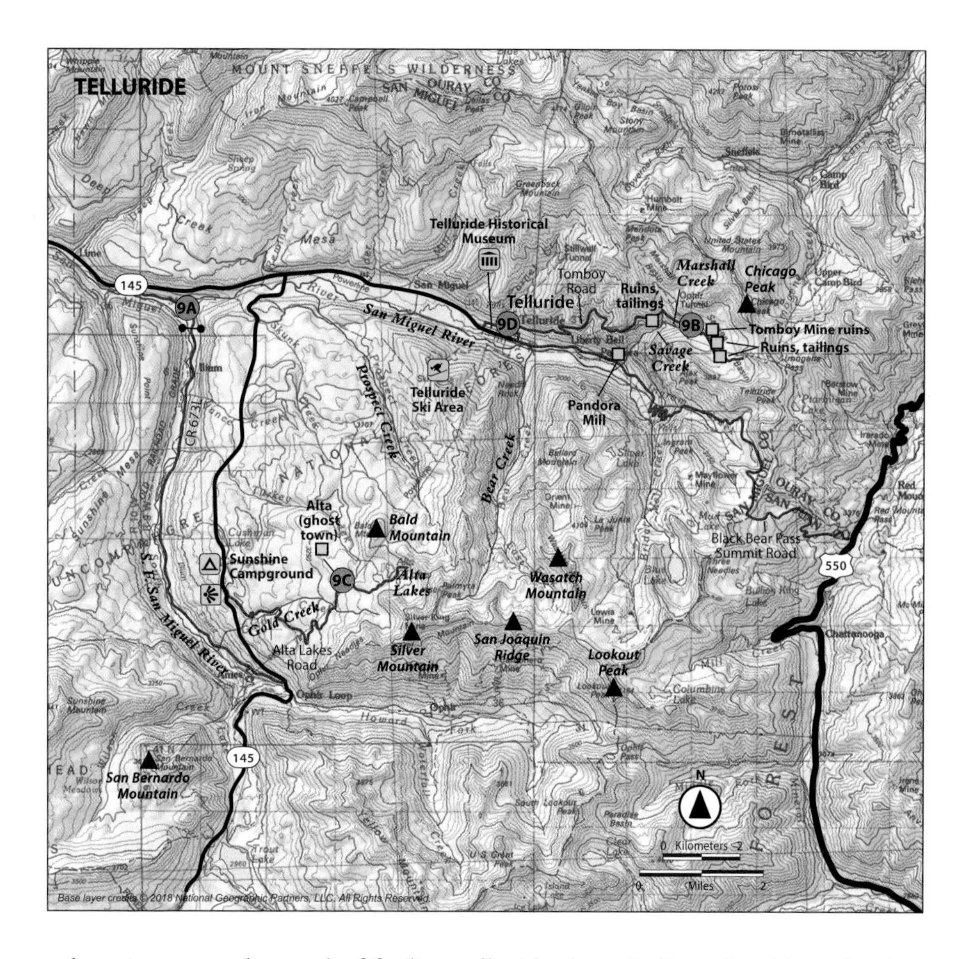

there is no actual record of finding tellurides here. Still, total gold production amounted to over 3 million ounces before 1959 (Koschmann and Bergendahl 1968, 116).

Site A offers access to both forks of the San Miguel River, but there is no bedrock to find. You'd be lucky to pan out a few colors at most unless you can start a really big hole, which you will only have to fill back in. The Tomboy Mine at Site B is a rare high-elevation water source for much of the year; this is Savage Creek, before it joins Marshall Creek. You should be able to get a nice sample of black sands with some pinpricks of gold; there are lots of traps, but bedrock takes some effort to find, and the water pours through here quickly and scours out the concentrates. Site C is the Warren Placers, below Alta Lake. The road turns away from Gold Creek, and it's a nice spot for a sample. There are lots of photogenic ruins here, too, and some tailings to bang

Savage Creek drains the basin above Telluride and contains a few colors.

Gold display at the Telluride Historical Museum at 201 W. Gregory Ave.

a hammer on. Site D is the Telluride Historical Museum, which has excellent mineral displays and features a panning tub for the kids on weekends. Parking is a bit limited, but you can get a temporary pass for street parking from the museum.

10. Rico

Land type: Riverbank
County: Dolores
Elevation: 8,822 feet in downtown Rico
GPS: A - Atlantic Cable Mine: 37.69455, -108.03181
B - Rico Boy ruins: 37.68741, -108.03612
C - Burnett Gulch: 37.66311, -108.03754
Best season: Spring
Land manager: San Juan National Forest
Material: Fine gold, rare small flakes
Tools: Pan, sluice
Vehicle: 4WD suggested if leaving highway
Special attractions: "Downtown" Rico; Rico Historical Museum; Rico Hot Springs (37.70219, -108.03149)
Accommodations: Dispersed camping throughout area; Priest Gulch Campground and RV Park are notable.
Finding the site: From the intersection of CO 145 and CO 145S just west of Telluride, drive south on CO 145S for 24 miles. You'll see Rico come into view, and as you get closer to downtown, the coordinates for Site A are on the left. To reach Site B, turn left onto W. Soda Street, just past Site A. Take Soda for 2 blocks and turn left onto N. Picker Street. After 0.3 mile go straight, then turn left and cross the river. After about 0.1 mile, turn right onto W. Elder Street, go a block, then turn left and follow the main road to the ruins at the coordinates. To reach Site C, return to town and turn on Mantz Street to reach CO 145. Resume traveling south on CO 145 for 2.9 miles to a generous pullout at Scotch Creek, where there is an information sign about the toll road on the east and good access to the Dolores River to the west. The next 2.5 miles contain more good access and primitive camping along the river.

Prospecting

The upper Dolores River valley was first worked by fur trappers as early as 1832, but it wasn't until 1866 that the first gold was discovered by Colonel Nash and a team of eighteen Texan prospectors. The Ute Indians made prospecting untenable, so the miners departed. Three years later, prospectors led

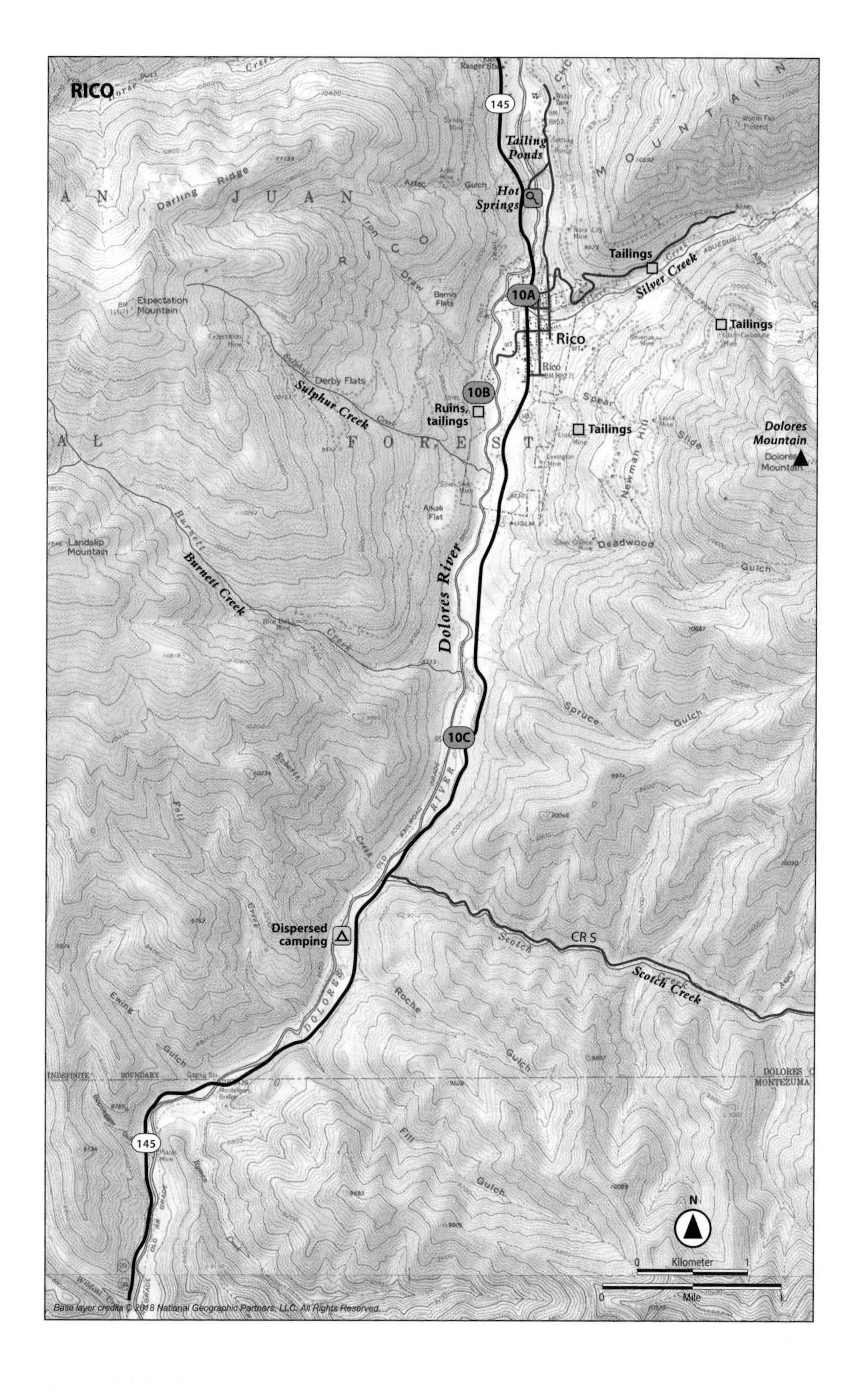
RICO
145
Tailing Ponds
Hot Springs
Tailings
Silver Creek
10A
Rico
Tailings
10B
Ruins, tailings
Tailings
Dolores Mountain
Sulphur Creek
Burnett Creek
Dolores River
10C
Dispersed camping
CR S
Scotch Creek
145
N
0
Kilometer
1
0
Mile
1
Base layer credits © 2018 National Geographic Partners, LLC. All Rights Reserved.

Picturesque mining ruins, interesting tailings, and a long stretch of the upper Dolores River will make you remember Rico fondly.

by Sheldon Shafer and Joseph Fearheiler traveled through the valley on their way to Montana and rediscovered Nash's lode workings, but more trouble with the Utes kept them from seriously working the area. After the Brunot Agreement was signed by the Utes in 1878, prospecting started again in the upper Dolores.

Nash's gold discovery became known as the Pioneer Lode and hosted the Shamrock and Potter Mines. The old town of Carbonate City sprung up when the Pioneer Mining District was established in 1876, and the district accelerated in the spring of 1879 when rich silver ledges were discovered on Blackhawk and Telescope Mountains. The town changed its name to Rico and was officially founded in 1879. The mountains soon hosted dozens of prospects and mines, and activity was booming. In October 1887 David Swickhimer struck the Swansea Vein at the Enterprise Mine, which reignited interest in the area.

The ore deposits around Rico contained large amounts of gold and significant silver veins, plus copper, lead, and zinc. Total production by the early 1970s is reported to be 83,000 ounces of gold, 14.5 million ounces of silver, 84,000 tons of lead, 83,000 tons of zinc, and 5,600 tons of copper (McKnight 1974).

Sites A and B are mostly photographic opportunities, with some interesting mineral samples at Site B. Site C offers primitive camping in the area, with lots of 4WD roads to check for good river access. The next 2.5 miles south have been open in the past, with no claim markers to dodge that we could see and plenty of dispersed camping right along the water. As usual, try to find the beginning of an inside bend or the upstream end of a small island, and get a deep hole started. Most of the gold here is fine, but if you can get deeper, you should be rewarded with larger pieces.

Farther down, Tenderfoot Gulch was busy back in the day, but it's mostly private now. There is also access to a good public area at the Road 436 bridge, farther still from Rico, but material is getting smaller and more scarce at this point.

The town of Rico has plenty of history to explore, and the city has put together a good walking tour. You'll find links to the tour and map at www.ricocolorado.org/vis/frame_vis.html. Hopefully the city planners will set aside a nice panning area open to the public as they continue their work.

11. Mancos River

Land type: Riverbank
County: Montezuma
Elevation: 9,165 feet at Golconda
GPS: A - Turn: 37.44239, -108.152703
B - Parking area: 37.44950, -108.14855
C - Golconda Crossing: 37.45637, -108.14703
Best season: Mid to late summer
Land manager: San Juan National Forest
Material: Fine gold, small flakes
Tools: Any
Vehicle: 4WD is suggested but not required. The road is in fairly good shape to the parking area.
Special attractions: Mesa Verde National Park
Accommodations: Primitive camping at Site B; developed campground at Mancos State Park. RV parking and motels at Mancos.
Finding the site: From Mancos on US 160, drive east 2.6 miles and turn left (north) onto CR 44. Drive 6.8 miles, then continue straight onto CR 566; avoid turning left onto CR 331. After 1.2 miles stay left on CR 566, and drive another 1.8 miles. The coordinates for Site A are where you leave the main road and go left for about 0.6 mile to a parking area and a gate. This is Site B. You'll have to walk about 1.1 miles to reach the river. The old camp of Golconda is near the ford.

Prospecting

The Mancos River area saw a brief boom in about 1887, with some limited production. Interest rose again during the 1930s, but also never amounted to much. There isn't a lot of information about the Golconda mining camp, so it must have been busy for just a few years, with but a few souls living in wooden buildings or tents. You'll be hard-pressed to find any foundations.

Nevertheless, the West Mancos still sees limited activity, with a dredge operating below the falls near the ford in 2017. There were no claim markers up, and we panned a nice, big flake from the first sample near the crossing. Most of the rest of the pans showed persistent colors, but very fine.

This is mainly an ATV area and the road is very rough after the parking area, so the gate makes sense. We speculated about how the crew got that

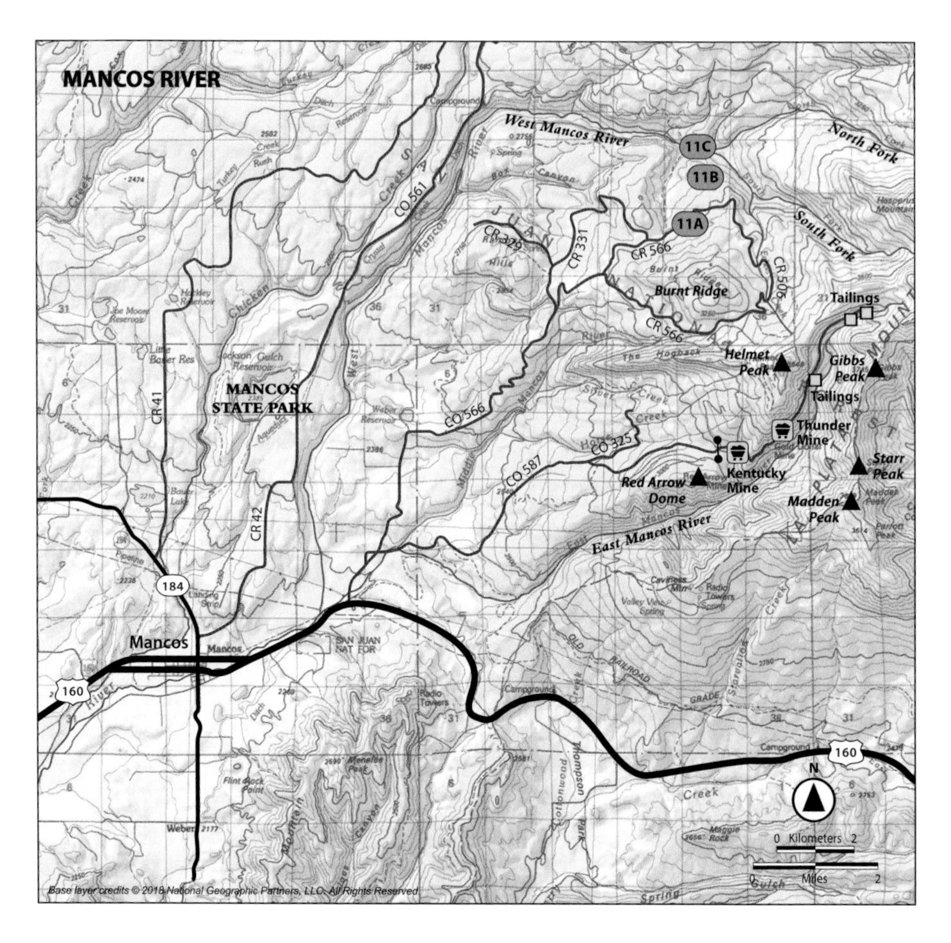

dredge in place, but they weren't around to ask. Maybe they got a key. You can also reach this area from the north, along more ATV trails, but the southern approach is a direct route.

Lower down, there is a lot of private land to dodge, so we gave up on that goal. The old Vulcan Placer, probably around 37.40714, -108.24771, is dried up due to the diversion of the West Mancos into a pipeline.

The East Mancos River is a different story—much harder to get to, but lots more activity. The lower stretch of the East Mancos is all private land, which is too bad. Farther up, the story is the same—posted and fenced off. There is a gate at 37.39253, -108.14546, blocking access to the Red Arrow Mine, which was a significant lode deposit that produced excellent wire gold and crystal gold specimens. It operated from 1933 to 1958, producing gold, silver, copper, lead, and zinc. Since access is blocked, go straight on CR 567 as

Dredgers were scouring around a big boulder at the base of the falls in 2017.

far as you dare; many prospectors walk from here to save their rig from getting beat up. The abandoned Kentucky Mine is at 37.39349, -108.13929, about 0.3 mile beyond the gate and 0.1 mile south of the road. There are some tailings there to hammer on.

Next is the Thunder Mine, above the road at 37.39696, -108.12782. The road is very rough and slow through here, with limited turnarounds, but you'll continue past more old workings and prospects, so bring a good hammer and break up any colorful rocks. The East Mancos is far below, and with the road at 11,000 feet, you'll need to be in good physical shape if you want to hike down (and back up). If you are making this trek, consider downloading the document "Geology and Ore Deposits of the La Plata District, Colorado," USGS Professional Paper 219 (Eckel, Williams, and Galbraith 1949).

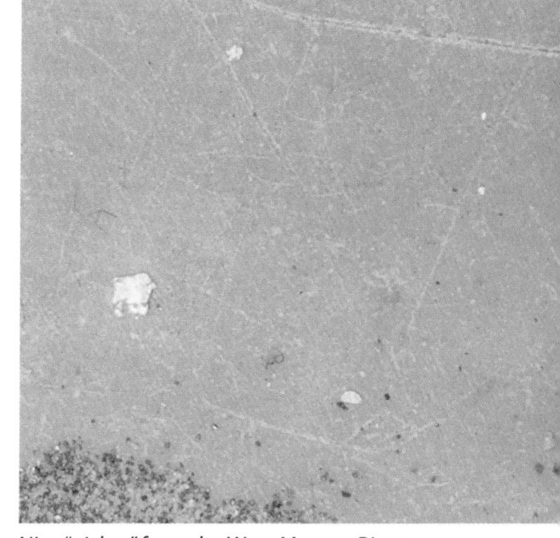

Nice "picker" from the West Mancos River.

12. La Plata

Land type: River canyon, mountain pass
County: La Plata
Elevation: 8,765 feet at Site A; 11,435 feet at Site E
GPS: A - La Plata City Campground: 37.39315, -108.06904
B - Basin Creek: 37.42214, -108.04762
C - Darby Campground: 37.42669, -108.04213
D - Columbus Creek: 37.44212, -108.03151
E - Cumberland Mill: 37.44641, -108.01282
Best season: Late summer; winter snow can come early.
Land manager: San Juan National Forest
Material: Fine gold, small flakes
Tools: Pan, sluice
Vehicle: Any for Sites A through D; 4WD suggested for Site E
Special attractions: Cumberland Basin
Accommodations: Developed campgrounds all along the river. Bay City Campground is free; Kroeger and Snowslide Campgrounds charge a fee, but only Kroeger has water. Motels and RV parking in Durango.
Finding the site: These sites are all in a line going up the La Plata River canyon. Start your mileage at US 160, just west of the junction with CO 140. The turn is about 11 miles west of Durango, or about 24.5 miles east of Mancos. Turn north onto CR 124 and drive about 7.8 miles, then turn into La Plata City Campground. There are some good information kiosks there. To reach Basin Creek, drive another 2.6 miles up the road; this is Site B. Site C is Darby Campground, about 0.5 mile farther. Site D is about 1.3 miles up, where Columbus Creek enters the La Plata River. Finally, Site E is 1.6 miles up the mountain, and the road gets more narrow and bumpier.

Prospecting

The town of Mayday was established in 1890 and served as a stop for the Rio Grande Southern Railroad, which greatly benefited the many mines above town. The Mayday Mine was the big producer, but it's on private land. The town grew to about 200 people and was even named the county seat in 1876, before Durango was founded and the county seat moved. There were dozens

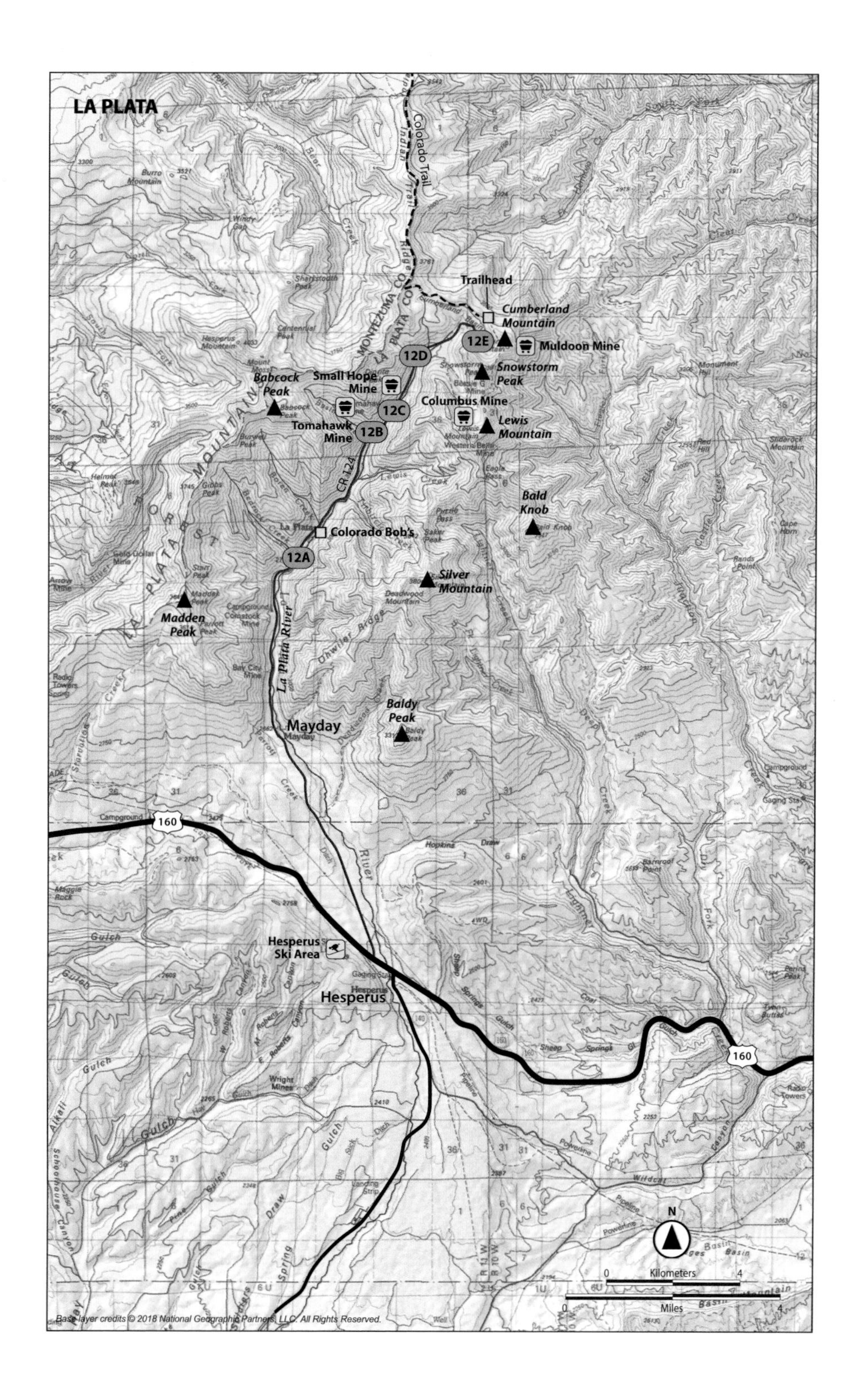

LA PLATA
Colorado Trail
Trailhead
Cumberland Mountain
12E
Muldoon Mine
12D
Snowstorm Peak
Small Hope Mine
Babcock Peak
12C
Columbus Mine
Lewis Mountain
Tomahawk Mine
12B
LA PLATA MOUNTAINS
MONTEZUMA CO.
LA PLATA CO.
CR-124
Bald Knob
Colorado Bob's
12A
Silver Mountain
Madden Peak
La Plata River
Baldy Peak
Mayday
160
Hesperus Ski Area
Hesperus
N
Kilometers
Miles
0
4
Base layer credits © 2018 National Geographic Partners, LLC. All Rights Reserved.

If you can make the drive all the way up to the ruins at the Cumberland Mill, it's worth the effort.

of mines and prospects in the mountains above the La Plata River, and you should still find plenty of color in the gravels if you can get a good hole going.

Even with five coordinates listed for this site, you can do plenty of additional exploring on your own. The best prospecting is far above Mayday, with multiple hard-rock mines that supplied gold to the gravels. This was a decent placer camp before it turned to lode mines, and while most of the flakes are small, if you look at them under a hand lens, they won't be highly rounded.

You can locate more access to the river, including at the multiple campgrounds, but we didn't find much color until we got up to the La Plata City area at Site A. Try your luck where the river makes a big elbow. Site B is near the mouth of Basin Creek, which is worth sampling. Farther up, get a picture of the big chimney, which is all that remains of the Gold King Mill. Basin Creek drains the Tomahawk Basin; the road up to those mines is about 0.3 mile past the coordinates for Site C. You might just want to hike it, as the road is miserable in places. Site D at Darby Campground gives you access to a natural chute that is moving too fast to trap any gold, so try below it. Up at the top of the mountain at Site E there are lots of mining artifacts and great vistas. Steer clear of private property, and stay away from the cabins. Leave the artifacts for the next traveler, and get plenty of pictures.

Here are some additional coordinates for more exploring: Tomahawk Mine: 37.42923, -108.05646; Small Hope Mine: 37.43376, -108.04145; Muldoon Mine: 37.44488, -108.00016; Columbus Mine: 37.42788, -108.01742.

One treat you'll find above Site D is Bob's Roadside Rock Shop, owned and operated by noted local rockhound and prospector "Colorado Bob" Ross. His website is www.bobsrocks4u.com, and he sets up his roadside rock shop most days at 37.39913, -108.06466 between Memorial Day and Labor Day. Bob was featured on an episode of the TV show *Prospectors* when the Busse family wanted some gold for a jewelry project. You're sure to find some good specimens for purchase at his shop. Bob is a wealth of local knowledge and extremely patient with all questions, and he's co-owner of the Bessie G Mine, a private claim located far up the road and the source of some outstanding mineral specimens. He will tell you to go as high as you can, sample every creek you cross, and bring a good hammer to collect ore samples.

13. Durango

Land type: Riverbank, city trails
County: La Plata
Elevation: 6,512 feet in downtown Durango
GPS: 37.25014, -107.87359
Best season: Late summer
Land manager: City of Durango
Material: Fine gold
Tools: Pan and hand tools mostly; no motorized equipment; sluice probably OK
Vehicle: Any
Special attractions: Multiple museums; charming downtown area
Accommodations: Hotels, motels, and RV parking in Durango. Junction Creek Campground northwest of town via CR 204.
Finding the site: Durango is located at the junction of US 160 and US 550. Drive south on US 160/550 for 1.2 miles, then turn right onto Frontage Road. After about 0.1 mile there is a left turn between the buildings to a large parking area where I took the coordinates.

Prospecting

Durango is a great little city to explore during the summer, with an outstanding rock and gem show in early July that you should try to include in your visit. We met up there with Wayne "Nugget Brain" Peterson, president of the Colorado chapter of WWATS, the World Wide Association of Treasure Seekers. Wayne had a brief moment of fame when he worked a short stint on Todd Hoffman's show *Gold Rush: Alaska*, operating machinery on Porcupine Creek for Fred Hurt.

Wayne told us there are multiple locales on the Animas River where you can pan, but after checking out spots in the city at Memorial Park, E. 29th Street, Rotary Park, and Schneider Park, we settled on the site below town at the East Animas Trail. If you're a local, you may have already sampled all those spots; if you're just visiting, I think my site is your best bet. There is plenty of parking and a restroom, and it's an easy hike to the big gravel bar forming there. When the water is low, the head of the gravel bar pops out, and there are plenty of big rocks to move around and start a big hole. Another good

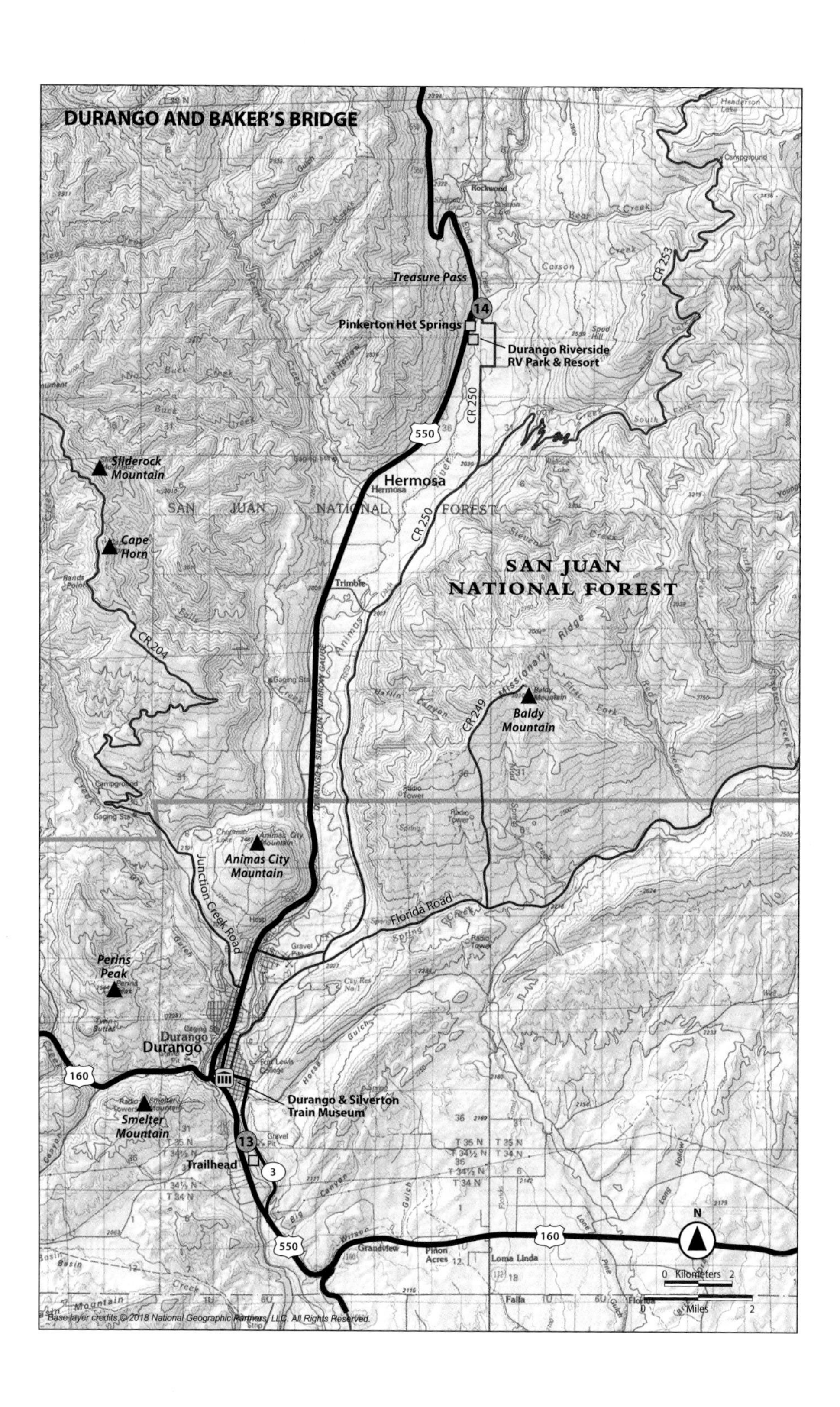
DURANGO AND BAKER'S BRIDGE
Treasure Pass
Pinkerton Hot Springs
Durango Riverside RV Park & Resort
Rockwood
Hermosa
SAN JUAN NATIONAL FOREST
Sliderock Mountain
Cape Horn
Trimble
CR 250
CR 204
CR 249
CR 253
550
Baldy Mountain
Animas City Mountain
Junction Creek Road
Florida Road
Perins Peak
Durango
160
Durango & Silverton Train Museum
Smelter Mountain
Trailhead
3
Grandview
Pinon Acres
Loma Linda
Falfa
0 Kilometers 2
0 Miles 2
N
Base layer credits © 2018 National Geographic Partners, LLC. All Rights Reserved.

It's not hard to see the iron staining from the 2015 Gold King Mine spill along the Animas River.

"shoulder" in the river can be found below the footbridge. The gold is very fine.

The rocks in the Animas River are stained yellowish orange due to the big spill upstream at the Gold King Mine in 2015. Alarmed by a growing leak in the mine's water retention structure, EPA engineers accidentally breached the plug, and an estimated 3 million gallons of iron-rich water cascaded into the Animas drainage. With time, the stain will become less noticeable, but it caused a mess throughout the drainage. Still, it shouldn't prevent you from panning out a few colors.

14. Baker's Bridge

Land type: Riverbank
County: Durango
Elevation: 6,784 feet at Baker's Bridge
GPS: 37.45892, -107.79958
Best season: Late summer
Land manager: San Juan National Forest
Material: Fine gold
Tools: Pan
Vehicle: Any
Special attractions: Durango & Silverton Narrow Gauge Railroad tour
Accommodations: Full facilities in Durango, including RV parking, hotels, and motels. Durango Riverside Resort & RV is right on the river, just 0.3 mile away. Dispersed camping on USFS land away from private dwellings. Camping at Haviland Lake and Hermosa Creek.
Finding the site: From the intersection of US 550 and US 160 in Durango, drive north on US 550 for 14.1 miles, then turn right to CR 250. After 0.2 mile cross Animas Springs Road and drive another 0.4 mile on CR 250. Park safely.

Prospecting

According to the historic plaque at this site, plus information provided by the nearby Durango RV Park, Captain Charles H. Baker brought a small party of fifteen prospectors to the Silverton area in 1860. Baker had already discovered gold in the territory, so he knew what to look for. His men built a bridge about 300 feet north of the present structure and founded Animas City about 1.5 miles north of present-day Durango. Unfortunately, there was probably more silver than gold in the pay dirt they were working, and the first winter was fierce. In 1861 Baker joined the Confederate forces fighting in the Civil War, which he survived, but he was subsequently killed by Indians when he returned to the San Juans.

To get a few good pans, you'll have to work your way downstream on the west side, past where swimmers flock in the summer. There are some good panning areas past the beach and rocks, where the outlet for two settling ponds empties into the river. Too much private land is here to do much

Work your way downriver to the right, where you can pan some decent gravels.

more; the actual placers were less than a mile north, working the area below the rapids. None of that is public land, so stay away. This area is "tolerated" as a local swimming hole, with bungee jumpers flocking to the bridge as well. Don't expect to do any serious digging here, but you should find lots of black sands and tiny amounts of gold, and you might see bits of native silver as well.

In the movie *Butch Cassidy and the Sundance Kid*, there's a famous scene where the posse has trapped Butch and Sundance at the top of a canyon above a river, and they are forced to jump into the water to escape. The jump was filmed here; the crew temporarily dammed the river to provide more flow, and camera placement made the jump seem much farther than it really was.

15. Silverton

Land type: City
County: San Juan
Elevation: 9,312 feet on Blair Street
GPS: A - Mining Heritage Museum: 37.81441, -107.66179 (1557 Greene St.)
B - Durango & Silverton Narrow Gauge Railroad Depot: 37.80789, -107.66301 (479 Main St.)
C - Swanee's Sluice: 37.81061, -107.66412 (Old Town Square on Blair Street)
Best season: Memorial Day through Labor Day
Land manager: City of Silverton
Material: Photographs at the museums; agates, amethyst, and pyrite at Site C.
Tools: Camera; pan provided at Swanee's
Vehicle: Any
Special attractions: Silverton Mineral & Gifts (1245 Greene St.)
Accommodations: Multiple developed campgrounds along Mineral Creek. Hotels, motels, hostels, inns, resorts, and RV parks in Silverton. Silverton Lakes Campground is right outside of town on Cement Street.
Finding the site: From Durango, drive north on US 550 for 48 miles, or if coming from the north, drive 23.4 miles from Ouray. Leave US 550 for Greene Street and travel through town. The museum is about 0.8 mile up at 1557 Greene St. This is Site A. To reach Site B, backtrack to E. 10th Street and turn southeast. The depot is at the end of the road. To reach Old Town Square on notorious Blair Street, turn off Greene Street at either 11th or 12th Street and turn onto Blair.

Prospecting

Silverton really comes to life in the summer, with tourists swarming Greene Street and Blair Street, popping in and out of the many shops that line the main thoroughfare. Many of the side streets are unpaved, and it's common to see rented ATVs zipping around. Several businesses in town rent 4WD vehicles, so you'll often encounter swarms of visitors headed in every direction from downtown Silverton.

The excellent historical museum at Site A has a great gift shop, superb exhibits, knowledgeable staff, and dozens of different books that detail the town's rich mining history. At 14,000 square feet, the museum includes the restored 1902 county jail, the archives for the San Juan County Historical

Two steam engines.

Kids can pan for "gems"—agates, amethyst, and pyrite mostly—at Swanee's in Old Town Square.

Society, and the Mining Heritage Center. Lots of rocks and minerals are on display, so you should start here to get a good feel for the geology of the area that the miners exploited. There are also great exhibits showing the tools of the early era, plus other historical artifacts. Plan your visit so that you have plenty of time to explore and don't feel rushed.

Site B is the famed depot and museum of the Durango & Silverton Narrow Gauge Railroad, in operation since 1882 and ranked as one of the top 10 scenic railroads in the world. The locomotive burns coal to generate steam, and it makes a great photo as it lumbers into the depot. There are daily rides between Durango and Silverton from May to October, steaming through the beautiful Animas River valley. The railroad also operates year-round smaller excursions.

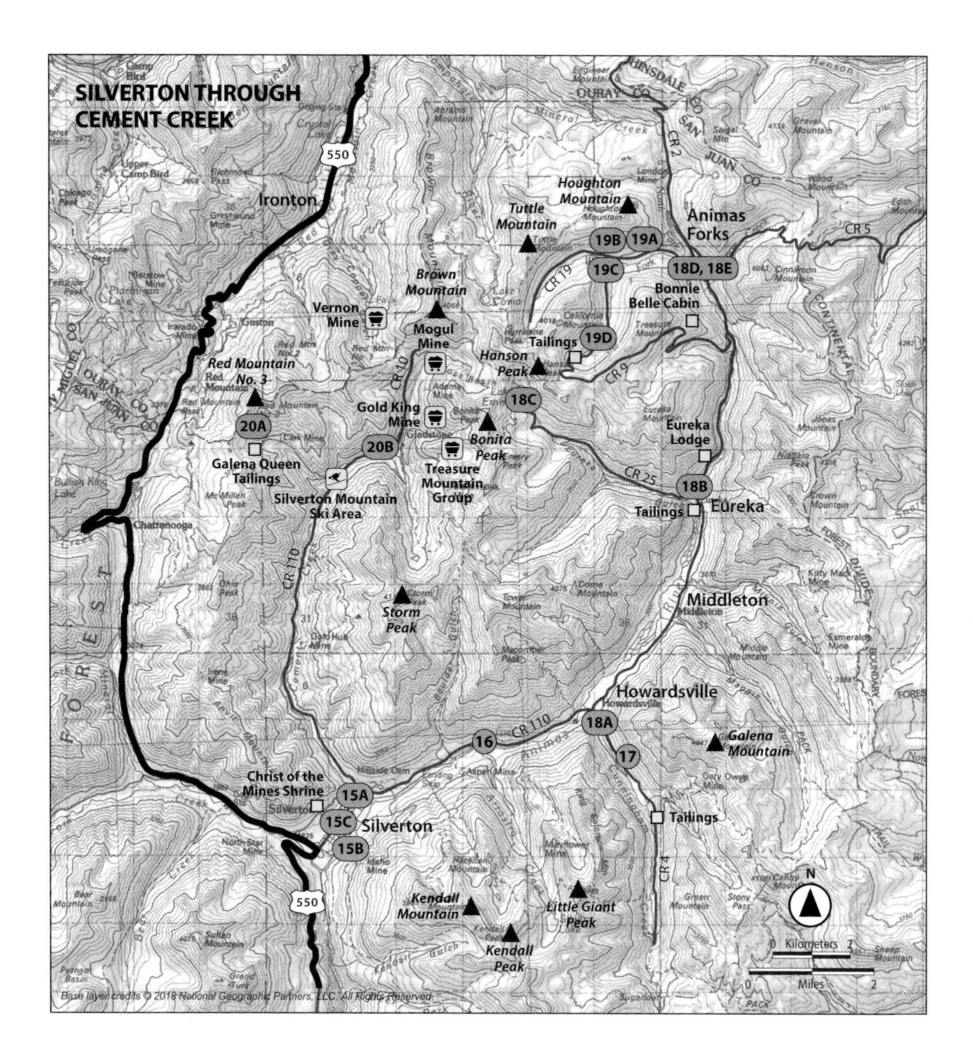

You can set up a round-trip or schedule bus-and-train combinations. There are also "specials" that may vary from year to year, but in the past have included such events as the Brew Train. Check www.durangotrain.com for more information.

Finally, check out the Old Town Square at Site C for some panning fun. The operators at Swanee's salt their pay dirt with semiprecious gems such as amethyst and pyrite, so the kids should come away with something for their treasure bag. The operators are knowledgeable and patient, and if you've never had a panning lesson before, this is a great place to learn how to do it.

Once you're done, there are two brewpubs in town: Avalanche Brewing and Golden Block Brewing. The pizza is good at both places, and so is the beer to wash it down.

16. Mayflower Gold Mill Tour

Land type: Lode mine
County: San Juan
Elevation: 9,662 feet at parking lot
GPS: 37.82862, -107.62611
Best season: June through Sep only. Closed outside of summer season.
Land manager: Private
Material: Photographs; pyrite and fine gold at the panning station
Tools: Camera; pan provided
Vehicle: Any
Special attractions: Nearby Arrastra Creek has small amounts of gold.
Accommodations: Multiple developed campgrounds along Mineral Creek. Hotels, motels, hostels, inns, resorts, and RV parks in Silverton. Silverton Lakes Campground is right outside of town on Cement Street.
Finding the site: From Silverton, drive east on Greene Street to the Mining Heritage Museum, then just past the museum, turn right onto CR 2. Drive about 2 miles, then turn left into the mill complex. Most visitors park down where the signs advise "flatlanders" to park and walk up to the gift shop. If you need to drop off closer to the entrance, that's fine.

Prospecting

The Mayflower Mill, also once known as the Shenandoah-Dives Mill, is a fantastic self-guided tour that takes you all the way through the mill, from the window where tram cars brought in ore from Arrastra Creek to the room where gold bricks were poured from liquid metal. The mill started in 1925, thus is in decent shape for a historical relic, and it was the last, and most advanced, of the great mills erected in the San Juan region.

Built at a cost of $375,000 in 1929 under the guidance of Charles A. Chase, the building used stout timbers of Oregon fir for framing and was completed in six months. By February 1930 it was in full operation, and it ran until 1991. The mill processed almost 10 million tons of sulfide ore and produced 1.9 million ounces of gold, 30 million ounces of silver, and 1 million tons of base metals such as lead and zinc.

Two processes are on display at the Mayflower Mill: gravity separation and selective flotation. Your handout will explain the different methods used at the

Imagine this shaker table lined with gold every day while the Mayflower Mill processed rich ore from Arrastra Gulch.

mill, and soon you'll end up at the shaker table, the assay office, and the pouring room. Because you can go at your own pace, you and your family should be able to enjoy the tour without rushing through. Once you complete the tour, you'll end up back at the gift shop, where you'll find books, ore samples, and more. Then take a turn at the panning station, which contains pyrite and tiny colors of gold.

A lot of preservation and renovation is still planned at the Mayflower. It is now owned and operated by the San Juan County Historical Society, and the mill gained status as a National Historic Landmark in 2000. Through grants and matching funds, the hope is that this unique operation will stand for years to come. Consider getting a Heritage Pass, which saves you money by combining the Mayflower Mill Tour, the Mining Heritage Center, and the Old Hundred Mine Tour into a package deal.

17. Old Hundred Gold Mine Tour

Land type: Mine
County: San Juan
Elevation: 10,029 feet
GPS: 37.82406, -107.58529
Best season: May to October only. Closed outside of summer season.
Land manager: Private
Material: Mine tour; agate, amethyst, and copper pellets at the panning trough.
Tools: Camera, warm clothes. Pan provided.
Vehicle: Any; 4WD strongly suggested for sightseeing. Bus connections available from the train depot in Silverton.
Special attractions: Eureka
Accommodations: None at site. Multiple developed campgrounds along Mineral Creek. Hotels, motels, hostels, inns, resorts, and RV parks in Silverton. Silverton Lakes Campground is right outside of Silverton on Cement Street.
Finding the site: From the north end of Greene Street in Silverton, turn right onto CR 2. Drive 4.2 miles to a right turn onto CR 4, and go 1 mile to the mine facilities.

Prospecting

The Old Hundred Gold Mine Tour is one of the best in Colorado. Riding an electric train into the mine is a treat during the hot summer months; the cool air of the mine is refreshing, and even occasional drops of groundwater from the mine's ceiling can cool you down. The drips do have an uncanny ability to find your exposed neck, however!

Young and old alike will enjoy putting on a hard hat and rain gear to prepare for the ride in. It also gets very noisy when the knowledgeable tour guides fire up the old machinery. You will learn about single-jacking, double-jacking, dry drills, wet drills, and "mucking" ore carts, and your guide will even explain the worst job in the mine—tending the "honey bucket." This was a small, portable toilet that ran on wheels, and it was the starting job for many a miner. We were told that it could also be a good way to "high-grade" valuable ores from the mine.

You'll see sulfide ores such as pyrite in the vein that the operators chased. This level served as an easier way to get ore out, rather than hoisting material up to the top. It also helped drain the mine. The tour is extremely safe

Kids will enjoy wearing hard hats and rain gear underground almost as much as they'll like the loud machinery in operation.

Be sure to leave time for a panning lesson at the Old Hundred Mine panning troughs.

and interesting, and your hour underground will go by quickly. No reservations are required, as tours run hourly. The gift shop has a good assortment of books, souvenirs, and ore samples.

The panning trough features some unusual perks. In addition to small pieces of gold, which turned out to be very elusive, the operators add small pellets and chunks of bright, shiny copper in the pay dirt, plus semiprecious gems such as agate and amethyst.

18. Animas Forks

Land type: Mine
County: San Juan
Elevation: 11,167 feet at Site E
GPS: A - Cunningham Creek: 37.83173, -107.59288
B - Eureka Mill: 37.88009, -107.56612
C - Sunnyside ruins: 37.90066, -107.61453
D - Gold Prince Mill: 37.92977, -107.56801
E - Columbus Mine: 37.93177, -107.57115
Best season: Mid to late summer only. After July 1 is usually safe.
Land manager: San Juan National Forest
Material: Fine gold, tailings
Tools: Pan mostly; sluicing possible
Vehicle: 4WD required beyond Site A. Roads are very rough and require good clearance.
Special attractions: Eureka ruins
Accommodations: Campground and lodge at Eureka. Full services in Silverton. Dispersed camping on USFS land, but avoid reclamation efforts, fragile alpine tundra, etc.
Finding the site: From the north end of Greene Street in Silverton, turn right onto CR 2. Drive 4.1 miles and take the turn to the Old Hundred Mine, then in 0.2 mile take the right turn onto CR 4. Drive another 0.1 mile to a large parking area above the creek. This is Site A. To reach Site B, return to CR 2 and resume traveling up the mountain. In 3 miles you can't miss the ruins at Eureka. This is Site B. The South Fork of the Animas River flows down Eureka Gulch just south of the ruins, and you can try panning for fines there. To reach Site C, drive a bit farther up CR 2 and after 0.3 mile turn left onto CR 25. Drive a total of 3.6 miles to the end of the road at Lake Emma; you can easily spot the Sunnyside Mine ruins for the last 0.8 mile. This is Site C; you can sample the South Fork between here and Eureka. To reach Site D, return to CR 2 and resume traveling north, up the mountain. Go 3.2 miles and turn left onto CR 9. Go 0.3 mile to the impressive ruins of the Gold Prince Mill and test the Animas River here, at Site E, or go another 0.3 mile and make your way to where the West Fork and the North Fork join to form the Animas River. See if you can tell which fork has more gold.

Prospecting

In the summer of 1860, a small group of prospectors worked their way up the Animas River. They found good placer ground as they moved up in elevation, but the limited season hindered their search in the mountains for the actual source of the gold. In 1874 the Bullion City Company struck paying ore at Cunningham Gulch, and Howardsville briefly jumped to life. It was erased by 1939. Middleton, so named because it lies between Eureka and Howardsville, sprang up in 1883, but it also quickly disappeared. Eureka was another short-lived site, getting a boost in 1896 when the railroad was extended from Silverton. The Sunnyside Mill closed in 1938.

The first log cabin in Animas City was built in 1873, and by 1876 there were thirty cabins, a hotel, a general store, a saloon, and a post office. In 1883 Animas City had about 450 people and boasted the highest-elevation printing press in the world. Soon the city thrived, with nearby mines extracting galena, copper, and silver. In 1884 a blizzard lasting twenty-three days dumped 25 feet of snow on the town, and residents used tunnels through the drifts to stay connected. The mines soon stopped paying, however, and avalanches were a constant threat. Even when the mines were going strong, many citizens spent their winters down the mountain in Silverton. In 1904 the Gold Prince Mill revived the town's fortunes, but it lasted only six years, and the parts were soon scavenged for the Sunnyside Mill in Eureka.

During the brief summer, Animas Forks can be pretty lively for a ghost town.

Now part of the Alpine Loop National Back Country Byway, linking Lake City, Ouray, and Silverton, Animas City has also benefited from being listed on the National Register of Historic Places. Work in 1997 and 1998 restored several buildings and identified rehab to do on many more; some of that work has been completed. The BLM now manages the area for tourism. There are nine buildings still standing, and you are free to enter them and take pictures. From Animas City, you can keep going on the Alpine Loop and drive your 4WD to Engineer Pass and then Lake City to the east, take Cinnamon Pass and come into Lake City from the south, or go on to Ouray to the north. Before continuing you should check with the BLM on road conditions, as avalanches, rain washouts, and other natural calamities are common and can block your route.

There are lots of opportunities for more exploring on the stretch of road from Silverton to Animas Forks, such as Arrastra Gulch, Maggie Gulch, Minnie Gulch, and Poughkeepsie Gulch, just to name a few. You'll see plenty of picturesque mining ruins, too—and some are for sale. Bring a good rock hammer to explore the tailings at any mine you find; look for brassy pyrite, dull gray galena, black sphalerite, and even some dull, gray silver ore. Most of the rocks are oxidized at their surface, so you'll want to break them apart to get to the fresh material.

At Site A, on Cunningham Creek below the Old Hundred Mine, you should be able to pan some colors by getting a hole started on the top of one of the many bends in the creek. At Site B, look for extensive tailings to the southwest of the Sunnyside ruins; if there is water in the creek there, you can try panning it, too. Just past Eureka, look for CR 25 headed up the hill to the west; you'll find lots of tailings piles and ruins up there at Site C, including galena and sphalerite, the ores of lead and zinc. Site D is at the Gold Prince Mill, near Animas City, with plenty of water and a good little bend above the bridge. You can easily reach two different forks of the Animas at Site E. Look for native silver in your concentrates all through here, as well as galena and sphalerite.

You should have no trouble collecting some interesting ore specimens at Eureka Gulch or up at the Sunnyside Mine.

19. Placer Gulch

Land type: Mountain gulches and creeks
County: San Juan
Elevation: 12,010 feet at Site D
GPS: A - Frisco Mill: 37.93274, -107.58051
B - West Fork: 37.93147, -107.58999
C - Placer Gulch: 37.92854, -107.59031
D - Sound Democrat Mill: 37.91203, -107.59499
Best season: July through Aug
Land manager: BLM and San Juan National Forest
Material: Fine gold, tailings
Tools: Pan mostly; sluicing possible some years
Vehicle: 4WD only
Special attractions: Animas Forks
Accommodations: Full services in Ouray and Silverton. Dispersed camping allowed on USFS land.
Finding the site: From Silverton, drive 12.4 miles up the mountain on CR 2 to the Columbus Mine at Animas Forks, and continue west on CR 9. Drive 0.6 mile to the ruins of the Frisco Mill, and if there's water, you can work your way down the hill to the West Fork of the Animas River and test it. This is Site A. To reach Site B, continue west on CR 9 for 0.6 mile through California Gulch, then turn right as CR 9 turns right. CR 19 will take you to California Pass and the headwaters of the Uncompahgre River. The fork in the road is Site B; it gives you good access to the West Fork of the Animas River again, and you should check it for colors. However, Placer Gulch empties into the West Fork below here, so you won't tap into that until you stay on CR 9 and drive at least another 0.2 mile. You'll have to hike about a tenth of a mile down to the water, but the coordinates at Site C at least offer a place to pull over. Otherwise, continue up to the Sound Democrat Mill at Site D.

Prospecting

The buildings above Animas Forks testify to the hardships miners are willing to endure in their pursuit of gold. At 12,000 feet elevation, the summers were short, transportation costs were very high, and finding workers was a big challenge. According to www.historycolorado.org, the Frisco-Bagley Mill

The Frisco-Bagley Mill was threatened by encroaching tailings and picked over by wood scavengers, but the foundation is actually quite sound.

and tunnel are part of the old Gorilla Mining Claim, built in 1912. The site includes the 150-ton reduction mill, a mine portal, and cement foundations for the compressors. This massive post-and-beam construction project was actually prefabricated. Each piece was precut, pre-fit, and coded with numbers and letters before being shipped to the site for assembly. The tunnel was driven into solid rock and has no timber framing.

Work there began in 1877, with the main portal started in 1904 and completed in 1911, after consolidating most of the claims in the area. A mining camp sprung up around the mill, but there are few remains. The mill has been a popular stop for tourists on the Alpine Loop for years. Sadly, artifact collectors have scavenged the area heavily, and the mill building itself was threatened with destruction in the 1970s. Much of the outside wood was removed for sale, plus metal scrap was torn down before the owners were able to stop the destruction. The Frisco Mill property is now for sale.

Mines up here had gold, silver, lead, copper, zinc, and more. The geology is complex but is part of the same Miocene volcanic caldera complex that includes the major mining districts of Telluride, Red Mountain, and Silverton. Placer Gulch was never a big producer, and the name is a bit misleading—there aren't any major tailings here, but there is water. After getting some pictures at Site A, you can pan a few colors in the West Fork of the Animas River

The Sound Democrat Mill is still standing near the head of Placer Gulch.

if you find a good trap at Site B. Keep going up Placer Gulch and prepare to bushwhack across the arctic tundra to the creek if you want to sample Site C. The walk is wet, as the tundra acts like a sponge. We snagged a couple of tiny colors from here, but the real treat was above the Sound Democrat Mill. The rocks are full of sphalerite, galena, quartz, pyrite, and silver ore. You can bang on the outcrops along the road down to the mill ruins and pan the creek on some obvious bends.

The Sound Democrat Mill was built in 1905–1906 and remodeled in 1909, and is a typical amalgamation and stamp mill to treat gold and silver-lead ores. Joseph T. Terry operated the mill; he was the son of John H. Terry, a prominent miner who left Ohio in 1859 as part of the Pikes Peak stampede. John Terry acquired an interest in the much more profitable Sunnyside Mine in 1889, and after John Terry's death, Joe Terry focused his attention away from the Sound Democrat. According to reports, the Sound Democrat may have produced as little as 200 tons of concentrate.

If snow levels permit, you can loop up and around Placer Gulch on CR 9 and after 2.3 miles reach the Golden Fleece Mine ruins on Picayune Gulch. Keep going down the mountain, and after about 1.3 miles you'll reach the ruins of the Sandiago Mine and the complex there. This was also apparently the site of the Blanchard Placer, at 37.91378, -107.56931. It's hard to imagine they had much water to work with. The main road, CR 2, is about 1.2 miles farther.

20. Cement Creek

Land type: Creek banks, tailings piles
County: San Juan
Elevation: 10,461 feet at Site A; 11,830 feet at Site B (Treasure Mountain Group)
GPS: A - Galena Queen ruins: 37.89187, -107.68827
B - Old bridge: 37.88946, -107.65381
Best season: July through early Sept should be safe; winter snows can come early some seasons.
Land manager: San Juan National Forest
Material: Fine gold; sulfide ores on the tailings at Site A
Tools: Pan, sluice
Vehicle: 4WD required for Site A
Special attractions: EPA reclamation project above Site A
Accommodations: Ouray and Silverton have motels, RV parking, and more. Developed campgrounds near Silverton; dispersed camping on USFS land.
Finding the site: From Silverton, drive through town on Greene Street, past the mining museum and the turn for Animas Forks, to CR 110 and turn left. Drive 5.7 miles on this decent gravel road to a left turn for FR 35. Unfortunately, FR 35 is not an easy road; 4WD is required here. The ruins of the Galena Queen Mine are about 1.9 miles up, at the end of the road. You will definitely want to walk around here. To reach site B, return to CR 110 and go about 0.5 mile farther up the mountain. A dirt road to the right, before the reclamation activity, leads to a big parking area and a rickety-looking wooden bridge across Cement Creek. This is Site B.

Prospecting

As you drive up Cement Creek, you'll notice the banks of the creek are a bright orange mess. This is the result of the Gold King Mine spill of 2015, which, as noted in the write-up for Site 13 (Durango), occurred when EPA inspectors inadvertently triggered a spill of 3 million gallons of iron-rich mine waste. The stain reached all the way to New Mexico at its height, but as we saw in Durango, the Animas is slowly cleaning itself up.

It's not hard to figure out the chemistry here; you can spot the tailings piles easily against the mountains in this valley because they stand out white and red against the brown hills. The chemical formula for galena, or lead

There are excellent ore samples around the Galena Queen ruins—especially galena.

sulfide, is PbS. Sphalerite, which is zinc sulfide, also contains iron, and its formula is (Zn,Fe)S. Pyrite is even richer in sulfur: FeS^2. It's easy to see there is always going to be sulfur-rich water in any of these flooded mines, and Cement Creek isn't the only creek in this area that stains its banks.

The Galena Queen Mine at Site A is full of good ore samples. Use the heft test: Pick up a rock and see if it feels heavier in your hand than you expected. Then crack it open with a big rock hammer. There is also pink rhodonite on the dumps here. Rhodonite is a manganese silicate—$(Mn,Fe,Mg,Ca)SiO^3$. The ratio of manganese, iron, magnesium, and calcium will vary, and these can be a nice pink. I logged the coordinates from a spot where I gathered a giant boulder of galena, which gave off a distinctive sulfur smell when I broke it apart. There are many more tailings piles within sight of this one, so spend some time here.

The EPA cleanup operation above Site B was going strong in 2017. Site B exploits the fact that the old tailings being covered up here were not thoroughly relieved of their gold during milling. You won't find much color in Cement Creek below here, but at this spot, you should easily find some small colors.

Interesting ore sample from the Galena Queen ruins, including pink rhodonite and brassy pyrite.

The road into the Gold King site is gated. Old maps show a complex of ruins at the Black Hawk Mine, but that's private property and also gated. There are many other old mines and ruins to explore with your 4WD above Site A. From the maps, CR 20A will take you to Corkscrew Gulch and then Ironton on US 550. CR 10 will take you to Lake Como, Poughkeepsie Gulch, and Animas Forks, according to the maps, but I can't vouch for those roads—although I can imagine how bad they are. Here are some other coordinates from Cement Creek drainage that I haven't been to, but might be worth the effort: Treasure Mountain Group: 37.89131, -107.63345; Mogul Mine: 37.91004, -107.63859;Vernon Mine: 37.91809, -107.65504.

21. South Mineral Creek

Land type: Creek, tailings
County: San Juan
Elevation: 9,499 feet at Site A; 10,637 feet at Site C
GPS: A - Kendall Campground: 37.81932, -107.71415
B - Sultan Campground: 37.82025, -107.72024
C - Bandora Group: 37.78626, -107.80122
Best season: July through Aug
Land manager: San Juan National Forest
Material: Fine gold
Tools: Pan, sluice, hammer
Vehicle: Any for Sites A and B; 4WD required for Site C.
Special attractions: Ice Lake
Accommodations: Multiple developed campgrounds along Mineral Creek. Hotels, motels, hostels, inns, resorts, and RV parks in Silverton.
Finding the site: From downtown Silverton, drive southwest on Greene Street for 0.5 mile to US 550. Swing north and drive 2 miles, then turn left onto CR 7/CR 585. Drive 0.7 mile to Kendall Campground Road and pick a spot near Mineral Creek. This is Site A. To reach Site B, drive farther up CR 585/CR 7 for 0.4 mile as it parallels South Mineral Creek. The coordinates are for the turn to the second campground, although Anvil Campground, on the other side of the main stem of Mineral Creek, is almost as good. To reach Site C, return to CR 585/CR 7 and drive 5.5 miles to an easy parking area just below the tailings pile and ruins.

Prospecting

South Mineral Creek was never a big placer producer, but the first two coordinates will access the main stem of the creek, and it carries good color. Site A at Kendall Campground offers good access to a point just below the confluence of Mineral Creek and South Mineral Creek. At Site A the water is discolored, but you should be able to pan out some small colors if you can find a good trap. At Site B your best bet is the west side, at Sultan Campground. Like Kendall and Anvil Campgrounds, it is a free site. (Golden Horn Campground and South Mineral Campground are both fairly expensive, at about $20 per

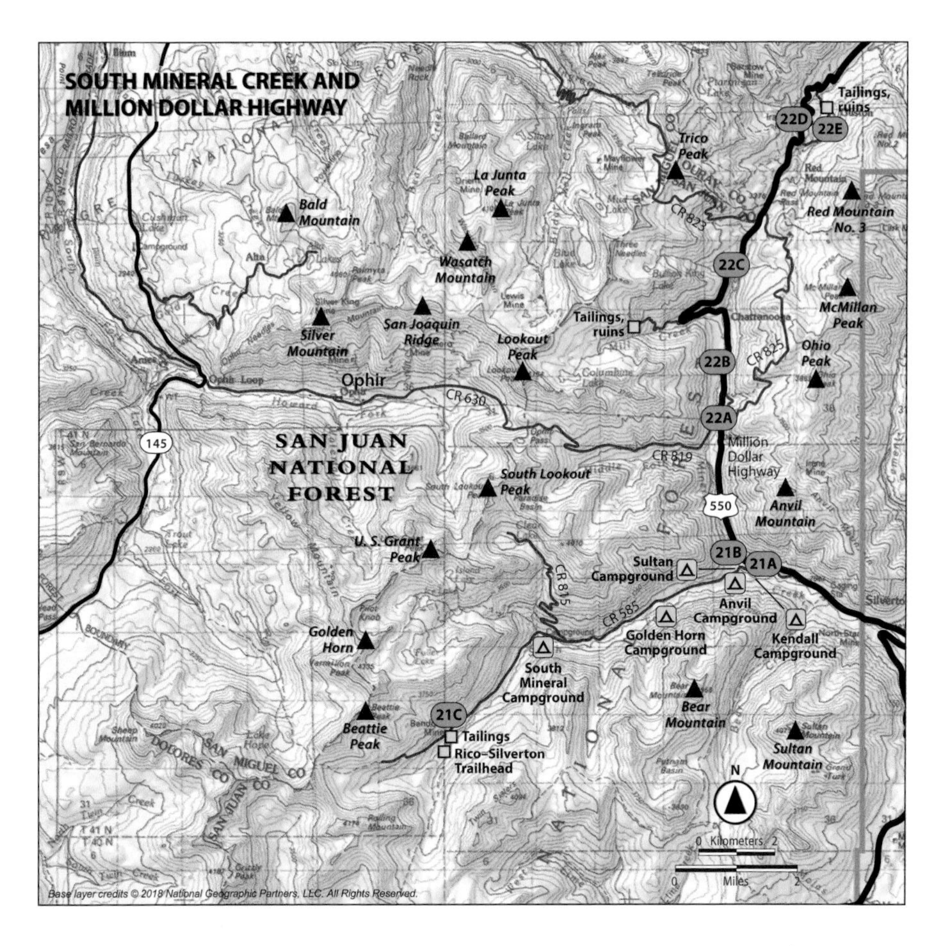

night.) There's better access to Mineral Creek from the west, and you're not up against the cliffs, so that's why I used it.

Site C is the Bandora Group of mines, with excellent tailings to swing a hammer. It's another typical Silverton-area mine, exploiting silver and base metals, with a little gold to add interest. We found good samples of pyrite and what looked like silver. Follow the debris up to the old wooden structure, and swing away at anything that looks interesting. The mineralization tends to show up best in rusted, oxidized samples, but if they're too rusty, you won't find anything fresh inside.

The Bandora Mine near the top of South Mineral Creek offers excellent tailings to collect ore specimens.

We sampled the various creeks (and culverts) leading up to the Bandora Group, but without much success. There are plenty of black sands, and what looked like lead and silver in the pan, but not much color in South Mineral Creek. But the main stem of Mineral Creek is a different story. You should be able to pan out some decent colors in each pan if you're sampling in a good spot, such as a natural trap. If you can get a hole going, you might do well—especially at Sultan Campground.

22. Million Dollar Highway

Land type: Creek banks, tailings piles, and ruins
Counties: San Juan and Ouray
Elevation: 10,151 feet at Site A; 10,637 feet at Site E
GPS: A - Mineral Creek: 37.85079, -107.72612
B - Ford: 37.86307, -107.72464
C - Silver Queen Mine: 37.88475, -107.72269
D - Yankee Girl overlook: 37.91533, -107.70116
E - Yankee Girl ruins: 37.91491, -107.69637
Best season: July through Aug; Sept for nice foliage colors
Land manager: San Juan National Forest
Material: Fine gold, small coarse flakes
Tools: Any (very little water); hammer for tailings
Vehicle: Any for road spots; 4WD suggested for exploring
Special attractions: Red Mountain Pass
Accommodations: Full services in Silverton and Ouray. Dispersed camping on USFS land away from private dwellings.
Finding the site: From Silverton, head north on the Million Dollar Highway (US 550) and drive 2.8 miles to the left turn for CR 679. Drive about 0.3 mile to the bridge; this is Site A. To reach Site B, backtrack to US 550 and resume traveling north. To reach Site C, return to US 550 and go north. After 1.9 miles FR 821 heads west along Mill Creek; there is some panning here and some mine ruins above you. This is optional; keep going north for 1.2 miles to the big pullout on the right. It's a short hike down the hill to the Silver Queen ruins, but at this altitude, make sure you're healthy enough for the hike back up. This is Site C. Keep going north for 2.9 miles to reach the Yankee Girl Mine and mill overlook at Site D. To reach more of the Yankee Girl ruins, drive farther north on US 550 for 1.4 miles and turn right onto FR 31. It will wind around for 1 mile, past the ruins of Guston and the Yankee Girl boardinghouse, to Site E. CR 31A is frequently washed out, but it takes you to the bottom of the slope where the Yankee Girl stands.

Prospecting

The Million Dollar Highway was first built as a toll road by Otto Mears in 1883, connecting Ouray to the growing mining camps at Ironton. A second

This bridge and ford spot at Site A on Mineral Creek yielded colors in every pan.

toll road extended over Red Mountain Pass to Silverton. By 1924 the road was rebuilt—a very expensive project, hence the name.

It's hard to say if "driving while prospecting" is a net plus. You'll certainly go slower than normal, spotting old tailings piles in the hills above you. But the temptation to get distracted could have every passenger in the car howling, so use discretion. This is a beautiful, and dangerous, road. It's narrow, with few shoulders in some places, and busy. Driving it when the snow is falling and ice covers the road would scare me silly, so I guess I'm a flatlander at heart.

Few mining districts were more important than the Sneffels–Red Mountain District. The Sneffels camp was at Imogene Basin, dominated by Camp Bird, while Red Mountain is at the top of the Million Dollar Highway. Geologists consider the Sneffels, Red Mountain, and Telluride Districts as a single entity, according to Koschmann and Bergendahl (1968, p. 107). They summarize work from various field investigations, especially W. S. Burbank, describing the area as a giant volcanic caldera: "The Sneffels and Telluride districts are in the exterior unit, on the northwest flank of the caldera; the Red Mountain area is in the northern part of the highly faulted outer ring." They go on to explain how the ore bodies developed: "Ores of the Red Mountain area are chimney deposits, which are vertical cylindrical bodies a few feet to a few tens of feet in diameter

Once you get on the back roads at Red Mountain, you'll find ruins and tailings around almost every corner.

in and near volcanic pipes filled with breccia, . . . porphyry, and rhyolite. The common ore minerals are pyrite, chalcopyrite, chalcocite, covellite, bornite, sphalerite, and galena. Some of the rare minerals include stromeyerite, enargite, and tennantite. Gold is associated with copper minerals" (108). Some chimneys contained ore as rich as 1,000 ounces of silver per ton.

In 1882 John Robinson discovered the Yankee Girl Mine, kicking off the stampede. About $30 million in silver, lead, zinc, copper, and gold flowed out of the district—close to $250 million at today's prices. After the Silver Panic of 1893, the district faded. The Idarado Mine operated until the 1970s, but the many boomtowns and camps are quiet now.

Just about every road up here heads for some kind of mine or tailings pile. Sites A and B are panning locales on Mineral Creek, but you should sample any stream, creek, or culvert you can find. Red Mountain Creek flows north, and is deeply stained bright yellow from the iron sulfide leaching out of the mines and tailings. It's worth sampling before it plunges into the canyon leading to Ouray, below Crystal Lake, but be cautious about private property and don't cross any fences. For a good map of hiking and 4WD touring opportunities, try http://ocs.fortlewis.edu/redmountainproject/heritagedaymap.pdf.

The rest of these sites require a hammer or camera. One caveat is the actual 10,000-acre Red Mountain Mining District, which is in private hands and is being stabilized and preserved. Nobody is going to complain if you pick up a souvenir rock from a tailings pile, but anything more, such as digging around and packing off artifacts, is strictly forbidden.

23. Upper Uncompahgre River

Land type: Riverbank, tailings
County: Ouray
Elevation: 11,647 feet at Site C
GPS: A - Switchback: 37.98702, -107.64767
B - Michael Breen Mine ruins: 37.97519, -107.63564
C - Access: 37.97014, -107.63053
Best season: July through early fall; road closed Dec through May.
Land manager: San Juan National Forest
Material: Fine gold; sulfides on tailings piles
Tools: Pan mostly; sluicing possible
Vehicle: 4WD above Site A
Special attractions: Ouray
Accommodations: Full services in Ouray. Dispersed camping allowed on USFS lands; developed campgrounds near Ouray.
Finding the site: From Ouray, drive south on US 550 for 3.9 miles. Turn left onto CR 18 and drive another 0.2 mile to the switchback, where you can park safely. Site A is to your right, and there is some bedrock on the west side of the water course. To reach Site B, continue another 1.4 miles up the mountain, but you'll see the ruins before you reach the coordinates. To reach Site C, drive about 0.5 mile farther up the mountain. CR 18A will take you farther up the Uncompahgre River, with more mines to explore in that direction, and leads to Poughkeepsie Gulch, while CR 18 leads to Engineer Pass, Animas Forks, and Lake City, with additional mine ruins to explore in that direction.

Prospecting

The headwaters of the upper Uncompahgre (pronounced Un-com-PAW-gray) River were prospected in 1874, and most of the mines are in the canyon walls. About 6,000 feet of rocks are exposed in the canyon, ranging from the Precambrian quartzite and shale of the Uncompahgre Formation to Tertiary rocks such as Telluride conglomerate, San Juan tuff, Silverton volcanics, and granodiorite intrusions. Ore bodies included three types of deposits: contact metamorphic deposits with a little gold; pyrite-rich base-metal deposits with silver and gold in tellurides, plus native gold; and lead-zinc deposits in quartz and barite.

The ruins of the load-out facility at the Michael Breen Mine are right on the road, and there are information kiosks explaining the operation.

Site A offers easy access to the river, and there are good colors in the creek here, along with plenty of traps and bedrock against the west side of the canyon to dig around. The Michael Breen Mine at Site B is a nice treat, with easy access to the ruins but a steep hike down to the water. It was located by William F. Sherman and Frederick Pitkin in 1874. The mine was driven into andesitic lavas, breccias, tuffs, and conglomerates, and it exploited fluorite veins for lead, silver, zinc, and copper mostly. Rock samples from the mine dumps contain small, clear quartz crystals. Mindat.org lists multiple minerals here, with bismuth, chalcopyrite, fluorite, galena, pyrite, rhodochrosite, sphalerite, tetrahedrite, and uraninite as your top targets. Some reclamation work has been done here, but there are still decent samples.

CR 18A keeps going to Poughkeepsie Gulch and California Gulch, but it's rough and you'll want to have the right vehicle. The road above Site C gets dicey as well, and worse, it is prone to washouts. One good resource for checking road conditions is from Lake City at www.lakecity.com/plan-your-colorado-trip/trail-road-conditions.

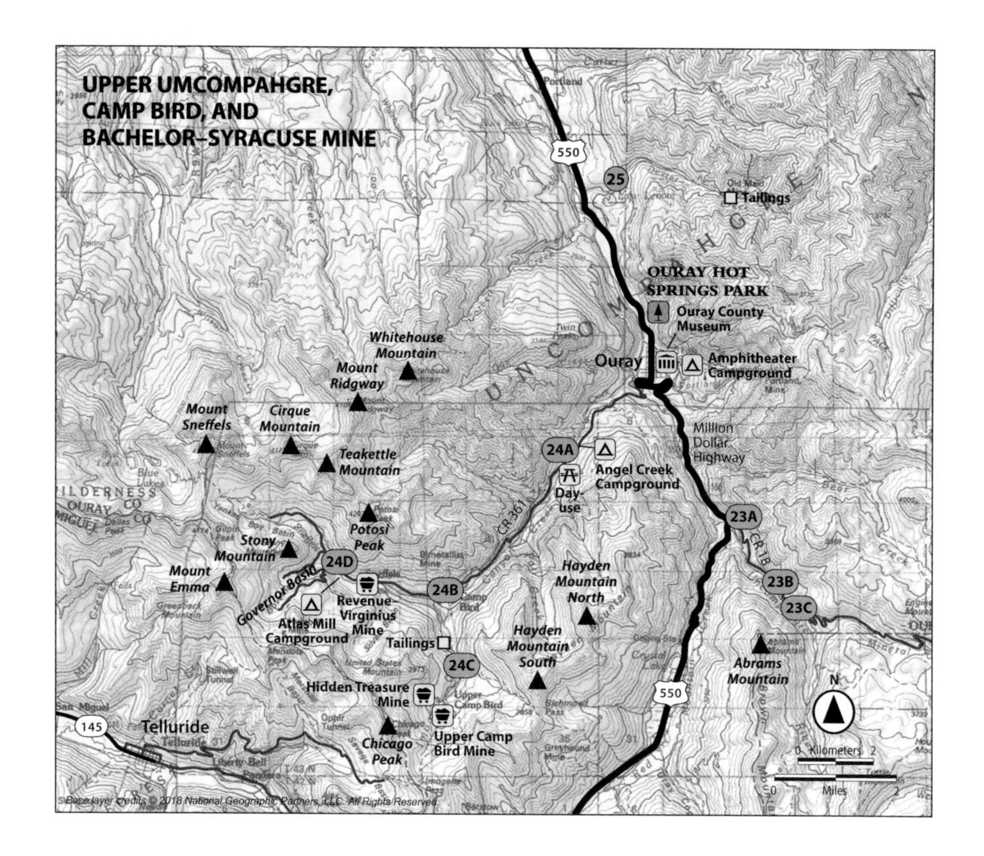

Don't be confused by the fact that there is a Mineral Creek flowing into the Uncompahgre River just above Site C, along CR 18. If you are able to continue up CR 18, such as to the San Juan Chief Mill at 37.96041, -107.59223, be sure to pull a test pan from Mineral Creek.

24. Camp Bird

Land type: Mountain creeks, mine ruins
County: Ouray
Elevation: 10,672 feet at Site C; 10,691 feet at Site D
GPS: A - Angel Creek day-use parking: 38.00251, -107.69378
B - Camp Bird: 37.97203, -107.72832
C - Yellow Rose Mine: 37.95697, -107.72399
D - Atlas Stamp Mill/Sneffels Creek: 37.97782, -107.75712
Best season: Late summer
Land manager: San Juan National Forest
Material: Fine gold, tailings piles
Tools: Pan, sluice
Vehicle: 4WD beyond Site A
Special attractions: Ouray County Historical Museum
Accommodations: Motels, hotels, inns, and vacation rentals in Ouray. RV parks and developed campgrounds in and around Ouray, especially along road to Camp Bird.
Finding the site: From Ouray, drive south on US 550/Main Street for 0.6 mile, then turn right onto Camp Bird Road/CR 361. Drive 2 miles up the mountain to the turn for Angel Creek Campground. The day-use parking area is Site A, with good access to Canyon Creek. To reach Site B, drive farther up Camp Bird Road/CR 361 for 2.8 miles to the junction with CR 26. Do not turn right onto CR 26 yet; stay left for 0.3 mile to a good parking area. This is Site B. To reach Site C from Camp Bird, continue up Imogene Pass Road/CR 361. To reach Site D, backtrack all the way to CR 26, then turn left onto CR 26/Yankee Boy Basin Road. Drive 2 miles along Sneffels Creek to the Atlas Mill ruins and primitive campground.

Prospecting

In 1896 Thomas Walsh discovered gold in Imogene Basin, about 2 miles from Camp Bird. There are some old buildings left at the Camp Bird mill site, with extensive reclamation under way there. It's mostly off-limits now, so your best bet is to keep going—in just about any direction.

Site A is at the lower part of Angel Creek Campground, which sits above Canyon Creek. The coordinates are for an easy day-use parking area where Angel Creek enters Canyon Creek. There are numerous traps and big rocks

The ruins of the Atlas Mill are crumbling fast.

here where you should get some nice colors. Thistledown Campground offered good access to Canyon Creek, and would be another good base camp to thoroughly explore the area. Site B is a good view of Camp Bird Mine, where Sneffels Creek comes in from the west.

If you continue up CR 361, also known as Imogene Pass Road, which is steep, rocky, and nasty in places (perfect for that 4WD you rented in Ouray!), you'll reach the Yellow Rose Mine. This is Site C, and it's worth the drive. The creek is easy to check for colors, and there are rocks in tailings piles all along the road as you get close, such as at 37.96239, -107.72814. The Upper Camp Bird Mine (37.94864, -107.73111) and the Hidden Treasure Mine (37.95007, -107.73241) are about 1 more mile up Imogene Pass Road.

This specimen from near the Yellow Rose Mine above Camp Bird was a treat to break apart with a 4-pound hammer. It contains pyrite, galena, and silver ore.

At some point you have to either turn around or end up back at the Tomboy Mine, headed for Telluride. For our purposes, turn around and head back to Camp Bird. If you want to take the rough shortcut on CR 26A, this is a scenic route to Site D before you reach Camp Bird. There are quartz veins along the road, along with collapsing cabins and waterfalls, and we found ourselves driving on a road that was doubling as a creek bed. This is basically the back way to the Revenue-Virginius Mine, which is active and closed to the public. Either way, the main road from Camp Bird is FR 853.1B, and it will take you to the ruins of the Atlas Mill at Site D. The coordinates are for a good parking spot, but you can easily access Sneffels Creek from here. This area has been reclaimed, so don't leave any traces after you pan out a few samples.

From Site D, you can either go west, to the Governor Basin area, or go right, up to Yankee Boy Basin. Either route offers plenty of picturesque ruins, ample tailings, and running water to get a few test pans. Governor Basin has active mining and reclamation, so use caution up there.

25. Bachelor-Syracuse Mine Tour

Land type: Mine tour
County: Ouray
Elevation: 8,044 feet
GPS: 38.05921, -107.67871
Best season: May through Oct only; closed during winter months.
Land manager: Private
Material: Photographs on mine tour; fine gold and occasional flakes at panning station
Tools: Camera; pan provided
Vehicle: Any
Special attractions: Ouray Hot Springs Park
Accommodations: Motels, hotels, inns, and vacation rentals in Ouray. RV parks and developed campgrounds in and around Ouray.
Finding the site: From Ouray, drive north on US 550 for 2.1 miles, then turn right onto CR 14/Dexter Creek Road. Drive 1.2 miles, then turn right onto Gold Mountain Trail. The mine is 0.2 mile ahead and has good parking.

Prospecting

In the early 1890s two unmarried miners, C. A. Armstrong and George R. Hurlburt, began digging in their free time, hoping to intersect the Al Mahdi Vein. Lacking any kind of dating app, they dubbed their workings the Bachelor Mine. There were no outcrops to guide them—they more or less worked on faith—and in three years they had only driven their adit about 500 feet, as they used hand tools. After adding Frank Sanders as an investor, they finally reached good silver and decent gold on what is now Gold Hill. They tunneled on the hill above the current adit and did well. The mine was a top producer in the Ouray District for years.

When the Bachelor Mine needed more funding, they pulled in a cash infusion from investors in Syracuse, New York, and created the Syracuse Tunnel. It did not meet up with the expected ore body, but it did offer an easier way to remove ore from the mine. It also helped with water removal and ventilation.

The mine stopped removing ore in the 1990s and lives on as a fun tour. After putting on a hard hat and listening to a quick safety lecture, you'll ride in the electric tram about 1,500 feet straight in along a narrow track. The tour

The panning material at the Bachelor-Syracuse Mine contains good gold, supplied by the "creek" that drains the mine.

guide will patiently walk you through the history of the mine and explain how the original miners used hand tools, then saw production take off once they used mechanical drills and dynamite. This tour isn't as noisy as some, and it's well-lit; when the guide turns off the lights, it can be unsettling!

The gold panning area is a real treat. You won't find gemstones or pyrite in a trough—this is a nice settling pond, and there are decent gold flakes. Unlike some panning areas where the gold pans are metal, with no riffles, the pans at the Bachelor-Syracuse are nice, modern Keene pans, just like the ones you'd want in your collection. I brought my own snuffer bottle. There is also a good gift shop and a restaurant on-site.

If you want to try some panning on your own, there is an easy spot and a hard spot. The easy spot is on the Uncompahgre River, just past the turn for the mine tour. Stay on US 550 about 0.5 mile, just past the sign for US Forest Service land. There is a good pullout for parking, and easy access to the river. The coordinates are 38.05499, -107.68986. The hard spot is up CR 14, about 1.3 miles past the turn for the mine tour. The road is very, very rough in places, but once you reach the old ford across the creek, park and look around for a good place to pull a sample. The coordinates are 38.06169, -107.66065. You can pan here or hike farther along Dexter Creek to the ruins of the Almadi Mine and the Old Main Mine, less than a mile away.

HONORABLE MENTIONS

The following locales did not make it into the book, but they're worth a visit if you're in the area.

Ophir

Near the pass from Silverton to Telluride, use Ophir Pass Road (FR 630) to access a charming little town along one of the forks of Mineral Creek. Lots of ruins, with limited access to the creek for much of the way. CR 819 accesses the creek at 37.84519, -107.74071 and might be worth checking. The Carbonero Mine at 37.86711, -107.81421 has tailings and ruins.

Poughkeepsie Gulch

Located between Ouray and Animas Forks, it looks promising if you can get to it. There are plenty of old reports of gold placer operations in this gulch. There is a big bend in the Uncompaghre River at 37.95699, -107.62645 that looks interesting.

Rifle Creek

Not a major producer, but worth checking if you pass through the area. The Sunshine Lode Mine was west of the present fish hatchery, around 39.69661, -107.70801. It produced zinc, lead, silver, gold, and copper, with manganese. Elk Creek might also be worth a shot; the Gray Eagle Mine produced gold, silver, zinc, and lead, and was located at around 39.67428, -107.52122. CR 241 would get you close enough to start sampling East Elk Creek.

Part II: Central Colorado

26. Lake City

Land type: Riverbank, mountain streams, high-elevation mine ruins
County: Hinsdale
Elevation: 12,800 feet at Engineer Pass
GPS: A - Day-use parking: 38.02169, -107.32821
B - Old bridge: 38.01387, -107.43254
C - Capitol City bridge: 38.00719, -107.46659
D - Bonanza Empire Chief Mine: 37.97389, -107.52264
E - Rose Cabin: 37.97629, -107.53831
Best season: Late summer
Land manager: Colorado Department of Transportation for Sites A and B; otherwise BLM
Material: Fine gold, small flakes; tailings piles
Tools: Pan mostly on Henson Creek; sluicing possible
Vehicle: 4WD recommended to Capitol City and required above Site C.
Special attractions: Engineer Pass
Accommodations: Full services in Lake City. RV parking at Lake Cristobal.
Finding the site: To reach Site A, drive to the south end of Lake City and turn right onto First Street, then left onto CR 20. Start up the mountain; after 0.6 mile there is an all-terrain-vehicle (ATV) day-use park, with good access to a small gravel bar on Henson Creek. To reach Site B, resume your travel up the mountain. In 6.4 miles you are driving alongside the creek, with multiple places to stop and test the panning. The coordinates are for a parking area near an old bridge. This road used to access the Pride of America Mine on Big Casino Gulch. Next up is the ghost town of Capitol City. You should see tailings along the road before you choose between going right on CR 24 or left on CR 20. There are mines along each road and colors in both forks. Stay left on the better road and head up CR 20 past the falls and the ruins of the Rose Lime Kiln at 37.97959, -107.49759 to the Bonanza Mine at Site D. It's about 4.4 miles. Finally, drive up the mountain another 0.8 mile, turn left onto the dirt road, and locate the ruins of Rose Cabin at Site E. It's an easy walk to the creek to sample the gravels. Since you made it this far, the top at Engineer Pass is an additional 4.6 miles of rough road. The old Frank Hough Mine, among others, is marked by extensive tailings.

The Ute-Ulay Mine complex is right along the road but is not open to the public. It is one of the oldest mine complexes in the Lake City District.

Prospecting

The gravels in the Lake Fork of the Gunnison River don't offer much in the way of color. If you want to try some of the fishing access spots maintained by the BLM north of Lake City, the bridge at 38.12664, -107.28944 is a good spot. There are amazing trout in this stretch of water, but not much color. Your best bet is up Henson Creek, although a lot of private land hinders access at the lower end of the drainage. The ATV area at Site A offers good access and a nice bend in the creek. After passing the Hard Tack Mine, photogenic ruins can be found at Henson, site of the Ute-Ulay Mine complex at 38.02069, -107.37759, plus good tailings to hammer on.

Site B offers good access to Henson Creek, and you should be able to pan colors, even though this was not a big gold-producing district. Look for galena samples on the tailings piles you encounter, plus sphalerite, the ore of zinc. Pyrite and chalcopyrite are also easy to spot. Silver and gold were by-products in this area, but you might find gray silver ore in the dumps.

Site C at Capitol City marks the bridge across the creek. You should be able to pull a quick sample here. There are tailings piles to the east along the creek. Steer clear of any dwellings and private land.

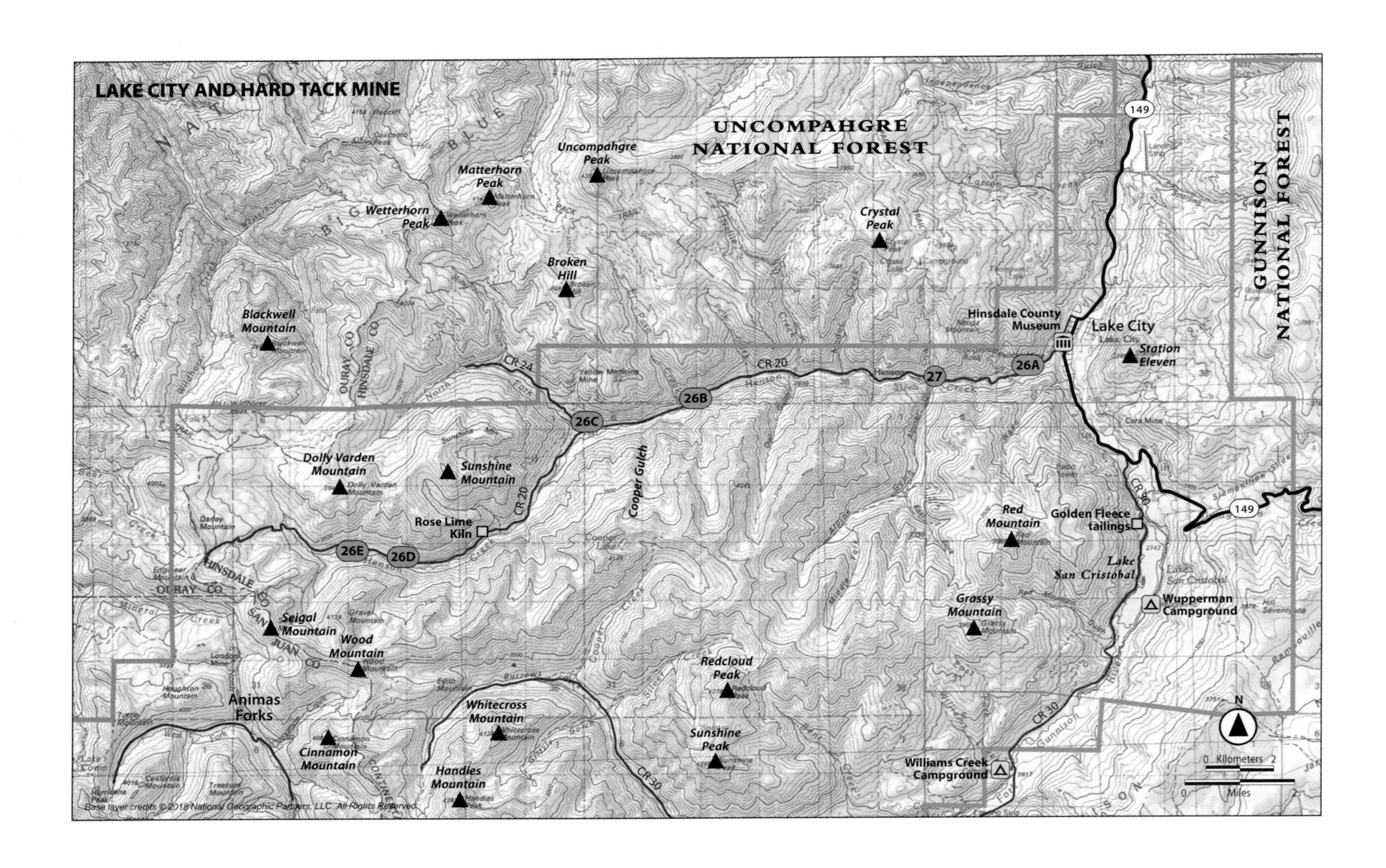

LAKE CITY AND HARD TACK MINE
UNCOMPAHGRE NATIONAL FOREST
GUNNISON NATIONAL FOREST
Uncompahgre Peak
Matterhorn Peak
Wetterhorn Peak
Crystal Peak
Broken Hill
Blackwell Mountain
Hinsdale County Museum
Lake City
Station Eleven
149
CR 24
CR 20
26A
27
26B
26C
26D
26E
Dolly Varden Mountain
Sunshine Mountain
Cooper Gulch
Rose Lime Kiln
Red Mountain
Golden Fleece tailings
Lake San Cristobal
Wupperman Campground
Grassy Mountain
Seigal Mountain
Wood Mountain
Animas Forks
Redcloud Peak
Whitecross Mountain
Sunshine Peak
Cinnamon Mountain
Handies Mountain
Williams Creek Campground
CR 30
HINSDALE CO
OURAY CO
SAN JUAN CO
N
0 Kilometers 2
0 Miles 2
Base layer credits © 2018 National Geographic Partners, LLC. All Rights Reserved.

View from the dumps above the ruined Bonanza Empire Chief Mine, looking up Henson Creek. You should be able to find pyrite up here easily.

The Bonanza Empire Chief Mine is a good example of the challenges of mining along Henson Creek. Good ore on the Bonanza Lode was initially discovered on July 4, 1885. In 1901 the Henson Creek Lead Mines Company issued stock worth $1.5 million and began building a mill at Site D. By 1906 they had most of the surface work completed and sold bonds to cover rising debts. About 3,700 feet of drifting underground showed promise, but by 1907 the mine was shut down. In the 1920s the mine saw a brief resurgence, but low commodity prices, rising operating costs, the beginning of the Great Depression, and a severe avalanche put an end to dreams of fabulous wealth. The Hindsdale County Historical Society stabilized the mill building in 2000, but eight years later an avalanche flattened the structure, and that's what you see here today. However, there are good ore specimens on the tailings at Site D. Hike up above the ruins to the flat area and pound away. Silver is hard to find, but galena is pretty easy.

Site E is the Rose Cabin, with scattered artifacts in the meadow. The deerflies chased us away from the creek pretty fast, but we did get some tiny colors from Henson Creek there.

You can continue all the way up Engineer Pass to Animas Forks if you have the right vehicle. Countless prospects are scratched into the mountains around the pass; the Hough Mine offers some good tailings to hammer. Meanwhile, back in Lake City, there are some low-elevation opportunities to collect samples. Voynick (2016, p. 40) mentions the Golden Fleece dumps along CO 30 at 37.98332, -107.29309. He also mentions quartz crystals at the mine dumps near Lake San Cristobal.

The Hinsdale County Museum at Second and Silver Street is worth a stop. It contains minerals from the collection of Harvey DuChene, a local geologist, and Paul Ramsey, who started working at age 14 in the local mines.

27. Hard Tack Mine Tour

Land type: Lode mine tour
County: Hinsdale
Elevation: 8,975 feet
GPS: 38.01955, -107.35794
Best season: June through Sept only. Closed during winter months.
Land manager: Private land
Material: Lode gold samples chipped from inside the mine; pyrite at the panning station
Tools: Camera; pan provided
Vehicle: Any
Special attractions: Hinsdale County Museum (corner of Second and Silver Streets in Lake City)
Accommodations: Full facilities in Lake City, including motels, hotels, inns, and vacation rentals
Finding the site: The Hard Tack Mine is located on the Alpine Loop National Back Country Byway, near the Hidden Treasure mill site and the ghost town of Treasure City. From Lake City, take CR 20 toward Engineer Pass for 2.5 miles and you will see Hard Tack Mine on the right side of the road.

Prospecting

The Hard Tack Mine Tour is run by Lake City residents George and Beth Hurd, along with their daughter Buffy. George was once a hard-rock miner, so he knows all the details behind the exhibits in the tour. He'll walk you through the old and new methods for drilling, explosives, and mucking, and the mannequins in the displays will bring the different facets to life. The walk is cool and level, and the best treat is the underground rock gallery. You'll see photographs from the old days, plus an excellent rock and gem display to show off the area's spectacular minerals. Don't make the mistake we made of sitting on the porch of the building outside the mine, waiting for someone to show up—go right into the mine!

There is a small panning tub outside the mine, where you can learn to pan from George if he's not conducting a tour.

The Hard Tack Mine was a tunnel to make it easier to move ore from the Hidden Treasure Mine directly above it. At one time a bustling little

Entrance to the Hard Tack Mine—head right in!

community surrounded the Hard Tack, with a mill, mine offices, boarding-house, and commissary—there was even a branch office of a bank. The miners sent their kids to school at the Ute-Ulay complex a mile up the mountain.

Voynick (2016) has an excellent article about this area in *Rock & Gem* magazine. He indicates that gold was first discovered on the Lake Fork in placer deposits by members of a US Army exploration team in 1842 near what is now Lake City. Since Ute Indians drove off any further prospecting parties, the next discovery wasn't until 1871, when the Ute and Ulay Veins were located. (The Utes pronounced the name of their chief Ouray as "you-lay," so that was the name the miners used.) But the two veins lay undeveloped, as the prospectors were trespassing at the time of their discovery.

In 1873 Alfred (or Alferd) Packer and five companions wintered over near Lake City, but by the spring only Packer emerged from their camp at Deadman Gulch. It soon came out that he had cannibalized his companions, and he served sixteen years in a territorial prison for his crimes. The Brunot Treaty convinced the Utes to cede about 4,500 square miles of land in the mineral belt. Prospectors soon swarmed the area, and Enos F. Hotchkiss quickly found the Hotchkiss Deposit, near the Ute-Ulay Veins.

By 1874 Lake City was a rough, bustling mining camp, and by 1875 its smelters and stamp mills were producing $500,000 in metals per year. Production reached $1 million by 1896, after the railroad reached town. The total for that year was 10,150 troy ounces of gold, 465,000 ounces of silver, 6 tons of copper, and 1,850 tons of lead. The Ute-Ulay and Golden Fleece deposits accounted for two-thirds of all production. The Ute-Ulay alone produced $12 million between 1874 and 1902, which extrapolates to about $280 million in today's dollars, according to Voynick.

Interestingly, Hotchkiss did not make out well with his find. He gave up, abandoning the workings, and when new owners renewed the filing, they called the property the Golden Fleece.

28. Creede Underground Mining Museum

Land type: Museum
County: Mineral
Elevation: 8,909 feet
GPS: 37.85728, -106.92779
Best season: Open year-round
Land manager: City of Creede
Material: Tour photographs; some color, native silver, and galena in the creek
Tools: Camera, pan
Vehicle: Any
Special attractions: Creede Historic Museum in the old Denver & Rio Grande depot (17 Main St.); Florissant Fossil Beds National Monument
Accommodations: Full services in Creede. No developed campgrounds, but dispersed camping OK if you avoid cabins, mines, etc.
Finding the site: From Creede, drive north on either Loma Avenue or Main Street to where they join. Continue north on what is now CR 503/West Willow Creek Road for 0.1 mile. You'll see the fire department tunneled into the rocks on the left, and then the signs for the museum. Parking is on the right.

Prospecting

Accounts vary, but according to Koschmann and Bergendahl (1968, p. 106), in 1883 J. C. MacKenzie and H. M. Bennet staked the Alpha claim at Sunnyside, on Miners Creek not far to the west of present-day Creede. The ores proved to be complex, and little production came from there initially.

In 1889 Nicholas Creede was a struggling prospector working the mountains above Wagon Wheel Gap (Dallas 1985, p. 51). Creede "picked aimlessly at an outcropping near where he had stopped for lunch. He examined the ore and gulped, 'Holy Moses!'" and thus located the Holy Moses Mine in Willow Creek Canyon. He sold part of the mine for $65,000 and then discovered the Ethel Mine, another valuable deposit. He then learned from two prospectors about valuable ore on Bachelor Mountain, and filed on a claim next to theirs. His claim became the Amethyst Mine. The technology for extracting silver had improved greatly by then, and the town mushroomed from 600 residents

There is a good panning spot on Willow Creek right across the street from the entrance to the museum.

to 10,000 by 1891. Several boom-and-bust cycles impacted Creede, and it attracted some of the most notorious figures in western US history, including Soapy Smith, Calamity Jane, Bat Masterson, Poker Alice, and Bob Ford, the man who killed Jesse James.

The US Geological Survey estimated that by 1966 the Creede District had produced 58 million ounces of silver, 150,000 ounces of gold, 112,000 tons of lead, 34,000 tons of zinc, and 2 million tons of copper. The last mine to shut down was the Bulldog Mine, operated by Homestake Mining until 1985.

Creede sits on a Tertiary volcanic caldera; according to the Colorado Geological Survey, the mineral belt here is just 25 square miles. The Creede Caldera started with a volcanic eruption and subsequent collapse about 26.5 million years ago. After the collapse the area underwent a complex history of hydrothermal circulation, dome resurgence, faulting, fractures, and fissures. The rhyolite host rock contains sphalerite, silver-rich galena, pyrite, rhodonite,

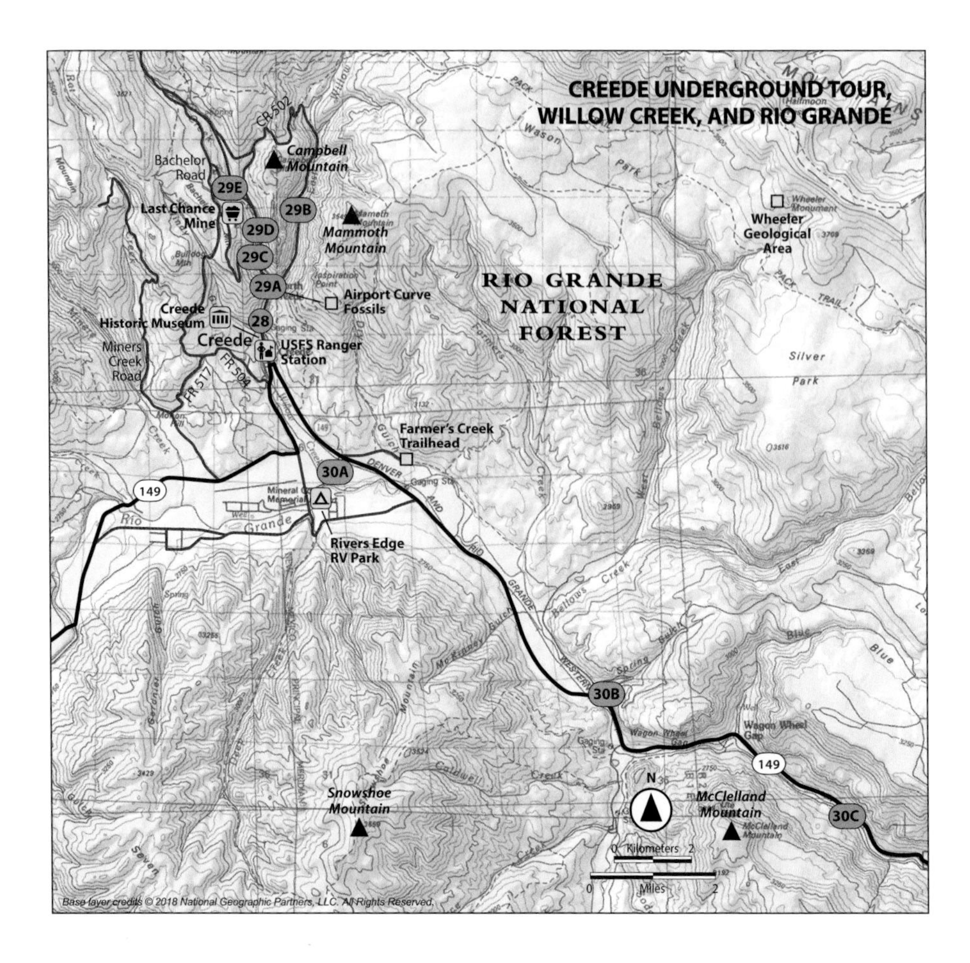

and chalcopyrite; gold values are spotty, sometimes rich and sometimes microscopic, associated with pyrite. The mineralization can occur in barite, quartz, fluorite, or chlorite, with the most famous occurring in amethyst.

The Creede Underground Mining Museum boasts excellent displays that show off mineral specimens collected from mines in the area. The self-guided walking tour includes an audio recording that explains underground mining techniques used in the district. There are pros and cons to listening to a recording: You can go at your own pace, but you can't ask follow-up questions. The museum isn't part of an actual mine, but it's laid out smartly and by the end you'll know the area's history and geology. The gift shop is excellent, with plenty of souvenirs, mineral specimens, clothing, and books.

There is no panning station at this museum, but the parking lot is adjacent to Willow Creek, and you can usually pan some color from the creek. You

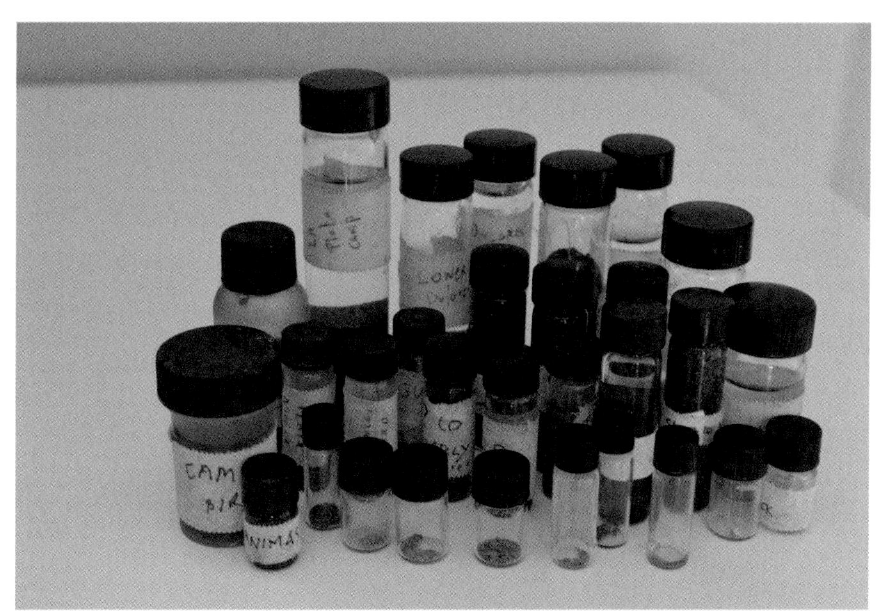

You're going to need a lot of vials to keep your samples under control, so there's always an excuse to visit the museum gift shop.

should also see native silver and galena in your pan, so the concentrates are interesting if you have a hand lens.

If you enjoy fossil hunting, there is an easy collecting locale at what the locals call the "Airport Curve," at the junction of CO 149 and Airport Road/ CR 806. This is the spot where the main road first swings north toward Creede, about 1 mile south of town. The GPS coordinates are 37.82942, -106.91796. The deposits are in ash beds about 13 million years old, similar to the famed Florissant Beds National Monument, and contain leaves from alder, poplar, pine, and fir. Some collectors have found fossil insects here by carefully splitting the fragile layers apart.

29. Willow Creek

Land type: Mountain creeks
County: Mineral
Elevation: 10,300 feet at Site E
GPS: A - Confluence of East and West Forks: 37.86485, -106.92536
B - Lower East Willow Creek: 37.88067, -106.91729
C - Commodore Mine: 37.86922, -106.92801
D - Lower West Willow Creek: 37.87429, -106.92831
E - Last Chance Mine: 37.88386, -106.93663
Best season: Late summer
Land manager: Rio Grande National Forest
Material: Fine gold, small flakes of silver
Tools: Pan; sluicing possible
Vehicle: Any, if careful; 4WD suggested
Special attractions: Bachelor Loop Historic Tour (17-mile drive)
Accommodations: Full services in Creede. No developed campgrounds, but dispersed camping OK if you avoid cabins, mines, etc.
Finding the site: From Creede, drive north on Main Street/CR 503 for 0.6 mile. Turn right onto West Willow Creek Road and drive 0.6 mile, then turn right and park. This is Site A. Follow East Willow Creek for 1.3 miles to Site B, then park safely. The creek is easy to access here. This road does continue on around the ridge, up and over to West Willow Creek, but it's a nasty 4x4 Jeep trail. It is much quicker to just turn around and return to Site A. Continue up the mountain on West Willow Creek Road for almost 0.4 mile to get some great pictures of the old mine structures at the Commodore Mine. Just above the Commodore you can easily access the creek again; we found a good spot at Site D, about 0.4 mile up. To reach the Last Chance Mine, continue up the mountain for 3.3 miles, then turn left onto Bachelor Road. Drive 1.4 miles, then keep left to continue on Last Chance Mine Road. The mine parking area is 0.6 mile ahead.

Prospecting

Panning gold in a major silver district is never easy—it takes patience and all of your best techniques. The locale at Site A is easy to reach, and you can practice some science by checking both forks of the creek and deciding for

Old workings of the Commodore Mine, above West Willow Creek.

yourself which one is better. Site B offers easy access to not only interesting tailings piles, but also a good spot on East Willow Creek. At Site C you should be able to grab some excellent photos of the old Commodore Mine, but the creek is too hard to reach here. If you continue up the mountain, Site D is easier to reach and there are good colors here, although small. There are plenty of big rocks to move, and you're sitting on a good bend in the creek.

Site E is the famed Last Chance Mine. It is worthy of a page all its own, except that it's more of a rockhounding site. You can pick over the tailings and collect your own specimens, or purchase material in their shop. Here's the description of the discovery of the Last Chance Mine from their website at www.lastchancemine.com:

> *The Last Chance was the last major strike in the Creede area. In 1890, prospector Theodore Renniger had unsuccessfully searched the area north of Creede. Feeling down on his luck, Renniger was about to give up. But when*

businessmen Ralph Granger and Erick Von Buddenbrock grubstaked him with $25 and three burros, the prospector decided to give it one more try. Knowing that this was his last chance to strike it rich, Renniger began searching Creede's steep ridges and rugged rhyolite canyons in August 1891. Things went wrong from the start; his burros wandered off, and when he finally caught up with them on the side of Bachelor Ridge, the stubborn animals refused to budge. Frustrated and weary Renniger decide to prospect right where he was. This time luck was with him, chipping away at a nearby rock formation, he uncovered a weathered outcrop of lead-silver mineralization in amethyst quartz that marked a north western extension of the Amethyst Vein. Renniger gave his discovery an appropriate name, the Last Chance. Renniger quietly asked Nicholas Creede to assess his strike. Stunned by its richness, Creede urged Renniger to immediately delineate, stake and register his claims. Once Renniger had done this, Creede quickly staked the adjacent lower ground and named his claim the Amethyst Lode.

These sites are all on the famed Bachelor Loop Historic Tour, a 17-mile bonanza of Creede history. What's listed here can only scratch the surface—there is a lot to explore, including nearby Miners Creek, which also has minor gold colors. Here's a link to the tour map: www.mininghistoryassociation.org/Meetings/Creede/Creede%20Map.jpg. Be sure to check in at the visitor center at 904 S. Main St. for additional information. You can find more info to plan your vacation at www.creede.com.

30. Rio Grande

Land type: Riverbank
County: Mineral
Elevation: 8,602 feet at Site A
GPS: A - Fish hatchery: 37.82264, -106.90872
B - Wagon Wheel Gap: 37.77584, -106.83288
C - Palisade Campground: 37.75177, -106.76608
Best season: Late summer
Land manager: Rio Grande National Forest
Material: Fine gold
Tools: Pan; sluicing and dredging probably not worthwhile
Vehicle: Any
Special attractions: Platoro Reservoir; Wheeler Geologic Area (tuff beds)
Accommodations: Full services in Creede. Multiple developed campgrounds along the river, both west and east of Creede. Mountain Views at Riversedge RV Park is right on the Rio Grande.
Finding the site: From Creede, drive south and east on CO 149S for 1.6 miles, then turn right onto the road to the fish hatchery. Drive about 0.4 mile and look for a big gravel area on the north bank of the river. To reach Site B, return to CO 149 and resume traveling east for 5.7 miles to a large parking area. Site C is the campground about 4.6 miles farther east.

Prospecting

On the way to Creede from Lake City, we stopped at one of the campgrounds along the river west of Creede and casually checked a few pans, not expecting anything. To our surprise we panned out a few colors—and this is *before* the best mineralization of the Creede District flows south into the Rio Grande, which is flowing east at this point. The locale at Site A offers access to the mouth of Willow Creek, which has meandered through the swampland south of Creede and empties into the river at two places within easy reach of here. The 12-acre Creede State Wildlife Area has boundaries you'll need to dodge; work your way up the east channel, above the bridge, or move upriver to the main stem of Willow Creek.

This stretch of the upper Rio Grande contains fine gold washed down from the Creede District mines.

At Site B you can get easy access to the Rio Grande, but the river isn't curving to your advantage here. Still, there is a nice beach area, and you should be able to pull a good sample. At Site C the Palisade Campground also offers good access, and the river makes a bit of a bend here by the day-use area on the west end of camp.

You'll probably want to check these spots when the water is lowest, later in summer. There is more access to the east, such as where FR 430 crosses the river, but you're going to want to be as close as possible to Creede to have a good chance at larger material. The gold is very fine here, and with so much black sand in your pan, you'll be challenged. Still, having looked out at the potholes and boulder traps of the Rio Grande at Big Bend National Park in Texas, and having walked along the Rio Grande at Ojo Caliente Hot Springs in New Mexico, it was a treat for me to pull some sample pans out of the headwaters of this great river and get to know it even better.

31. Summitville

Land type: Mountain creeks and rivers
County: Conejos
Elevation: 11,276 feet at Summitville
GPS: A - Little Annie Placers: 37.43199, -106.59961
B - Information kiosk: 37.43068, -106.59354
Best season: Late summer
Land manager: Rio Grande National Forest
Material: Fine gold
Tools: Pan; maybe a sluice in high runoff months
Vehicle: 4WD suggested; lots of washboards
Special attractions: Platoro Reservoir
Accommodations: Dispersed camping throughout area; developed campgrounds at Alamosa River and Platoro Reservoir. Platoro Cabin can be reserved online at www.recreation.gov. Gold Pan RV Park and Skyline Lodge in Platoro, but no gas.
Finding the site: From South Fork on CO 149, drive south on US 160 for 11 miles. At Demijohn Road, turn right and drive 17.3 miles to the Summitville area. Site A is the Little Annie Placers, below you on the right. To reach the information kiosk at Site B, drive another 0.2 mile, turn left, and continue 0.2 mile.

Prospecting

In early 1870 a small band of former Union Army soldiers prospected their way up the Rio Grande. They must have found enough color to keep going, because they apparently turned to follow the colors up the Alamosa River, then took the Wightman Fork to the top. Soon the boom was on, digging placer gold from what became the Little Annie Placers. Probably only $10,000 in placer gold came out of the creek, according to Koschmann and Bergendahl (1968, p. 111). Prospectors located the first lode mines in 1871, and by 1873 they staked the richest ground on the mountain. Production began in earnest by 1873. The surface ores were oxidized and easy to work with, but by 1887 they were exhausted after yielding about $2 million. In 1934 the best properties were consolidated, and by 1947 another $4 million came out of the district.

Production waned until 1984, when Summitville Consolidated Mining Company, Inc. (SCMCI) began a large cyanide heap-leaching project.

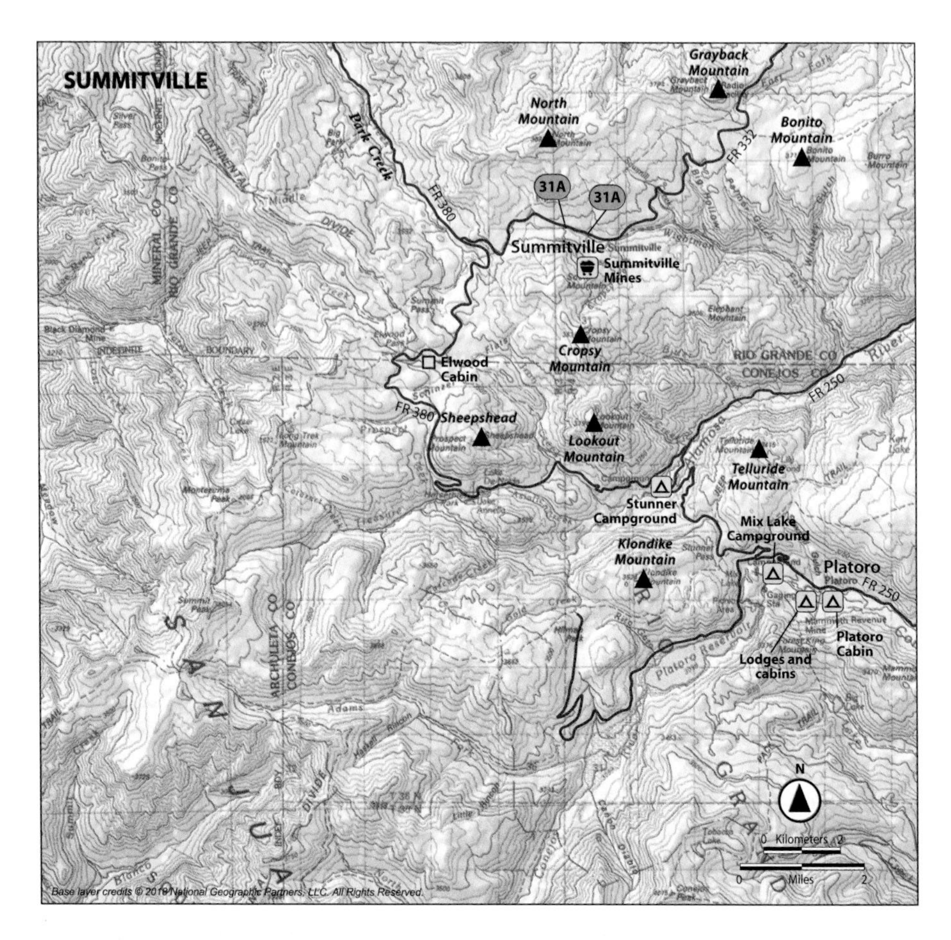

Unfortunately, in 1991 excess meltwater and a damaged pad liner resulted in the release of toxic metals and cyanide. The spill killed just about all aquatic life for 17 miles to Terrance Reservoir. SCMCI declared bankruptcy, the Environmental Protection Agency took over, and to date about $150 million has been spent fortifying the discharge ponds, erecting a treatment plant, and fixing the mess.

Ore bodies at Summitville occur in resistant quartz "pipes" and vein-like masses of vuggy quartz found in volcanic rocks such as rhyolite. Common minerals include pyrite, enargite, galena, and sphalerite.

On the way in to Summitville, we tested Park Creek but found only black sands. The Little Annie Placers offered up small colors, but while we were there, a fierce hailstorm blanketed the area, and it's not much fun panning when hail pellets are stinging every bit of exposed flesh. Since this locale is above the ongoing reclamation efforts, and you can't do any damage with a

The Little Annie Placers on Wightman Fork still contain fine gold. Steer clear of any reclamation work.

shovel and gold pan, you should be able to get a good sample. There are some excellent information kiosks at 37.43068, -106.59354 and plenty of old buildings to photograph. An access road used to follow Wightman Fork for quite a ways to the east, but it's gated now. What you *can* do is circle around to the Alamosa River and check the confluence of the Wightman Fork where it empties into the river at 37.40394, -106.52175.

There is a good reason why you might want to keep looking up here. This excerpt is from an article by Voynick (2014):

> *The Summitville gold boulder is Colorado's most remarkable chance discovery of gold in recent times. In 1975, ASARCO [American Smelting and Refining Company] geologists surveyed Summitville's low grade gold resource on South Mountain. The only free gold specimens found contained particles barely visible and no one expected to find much more. But on the afternoon of Oct. 3rd, an equipment operator noticed a large rock lying in full view just off the shoulder of a public road below the early mine ruins. The rock was gleaming back at him in the sunlight. He was shocked to find it laced with native gold. Returning excitedly to camp, he asked the project geologist if he wanted to see some gold. The geologist casually agreed, but when the operator said it would take two of them to put it in the pickup, he suspected a joke.*

Instead, the 141-pound float boulder was laced with about 350 ounces of gold. The operator was awarded a $21,000 "finder's fee," and the boulder is now on display at the Denver Museum of Nature & Science, valued at $500,000.

32. Bonanza

Land type: Mountain creeks
County: Saguache
Elevation: 10,774 feet at Site C; 9,751 feet at Site D
GPS: A - Kerber Creek: 38.28439, -106.13939
B - Exchequer town site: 38.30673, -106.14712
C - Rawley Mine: 38.32174, -106.12493
D - Bonanza Mine and Mill: 38.31295, -106.15058
Best season: Late summer for less snow; late spring for more water if you want to sluice
Land manager: Rio Grande National Forest
Material: Fine gold, tailings; copper ore
Tools: Pan, sluice, hammer
Vehicle: 4WD suggested; even the good roads are bumpy and have a lot of washboards.
Special attractions: Great Sand Dunes National Park
Accommodations: Full services in Salida. O'Haver Lake Campground isn't far away.
Finding the site: From about 0.3 mile north of Villa Grove on US 285, drive west on CR LL56 for 14.4 miles. Site A is along the creek at a turnout; park safely. Site B is farther north; drive 1.7 miles on CR LL56/Bonanza Road, then park near the old cabin. To reach Site C, continue up CR 47 QQ, past the first switchback, for 1 mile total, and look for a faint road heading back to the right, paralleling the creek. Drive about 1.2 miles to the sign for the Rawley Mine and turn right, down to the flats, where you can reach the creek and the mine ruins. To reach Site D, backtrack to CR LL56 and Site B, but turn right to go farther up the mountain. Drive about 0.1 mile and stay left; avoid CR 46AA. Stay on CR LL56 for 0.4 mile until you reach the ruins of the Bonanza Mine and Mill.

Prospecting

In 1880 prospectors found interesting mineralization along Kerber Creek, locating good silver-lead fissure veins. A small boom followed, and one of the enthusiastic miners proclaimed, "She's a bonanza, boys!" Soon the town of Bonanza was founded. Even former US president Ulysses S. Grant visited the area when it was predicted to be "the next Leadville." Dallas (1985, p. 30)

The ruins of the Bonanza Mill are surrounded by good mineral specimens in the tailings.

reports that Grant spent a week there and had his meals sent in from Salida. He offered to purchase the Bonanza Mine for $40,000 and tried to buy the Exchequer Mine for $160,000. Both offers were refused, probably to his benefit. The boom was short-lived, and by 1882 most of the citizens had moved on.

While the Bonanza Mine and Mill is the more photogenic of the big mines in the area, the Rawley Mine produced significant riches on its own. In 1911–1912 the Rawley Mine exploited an ore shoot with silver, lead, copper, and zinc. However, by the 1930s that property fell silent, too. In 1999 ASARCO completed a five-year remediation project at the Rawley Mine to prevent further damage from runoff.

Although similar to other volcanic caldera deposits in Colorado, Bonanza appears to be too far toward the edges of the great Colorado Mineral Belt to emerge as a mega-district. In all, only about 17,000 ounces of gold flowed from the area, mostly as a by-product of the 150 tons of silver produced there.

The placer locale at Site A is worth checking, and you should get colors. This spot is well south of the Bonanza city limits and didn't appear to have any issues with private property. If you take a right turn on First Avenue, you can go up Copper Gulch and see the ruins of the Empress Josephine Mine. The

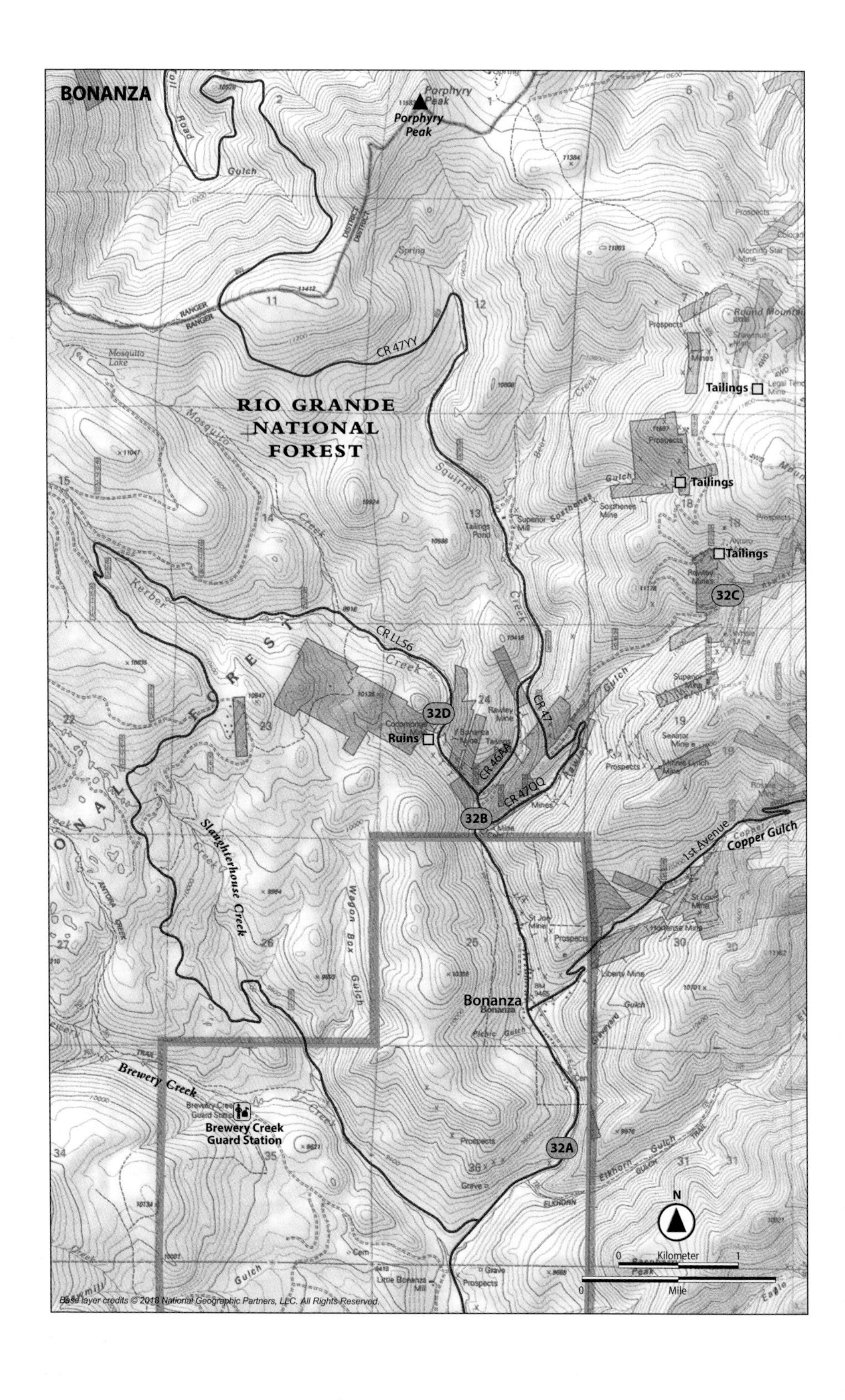
BONANZA
Porphyry Peak
RIO GRANDE NATIONAL FOREST
CR 47YY
CR LL56
CR 47
CR 46AA
CR 47QQ
32A
32B
32C
32D
Tailings
Ruins
Bonanza
1st Avenue
Copper Gulch
Slaughterhouse Creek
Brewery Creek
Brewery Creek Guard Station
N
Kilometer
Mile
Base layer credits © 2018 National Geographic Partners, LLC. All Rights Reserved.

Galena, minor pyrite, and possibly enargite in a hand sample near the mill.

creek in Copper Gulch is intermittent, so it might be hard to find enough water to pan. There are several old lode claims above you: the Queen City, Rosalie, Navajo Claims, and Senator Mine are all on the north slope.

The Exchequer "town site" at Site B offers you two panning options: Try Kerber Creek, the main drainage, or Rawley Gulch. Either contains colors, but Rawley Gulch is a better choice. Reclamation at the Rawley Mine has covered any chance of good ground to work, but lower down there are several places to reach the creek and pan. An avalanche apparently flattened many of the trees here a few years ago. Multiple old mines and prospects are located up this road, all the way to north of Round Mountain, but it is very rough going.

The Bonanza Mine is on Kerber Creek, and the main road follows Squirrel Creek. You should be able to find some good ore samples on the tailings piles in either direction; there is a lot of ore around the Bonanza Mill, with galena and pyrite both common. Geologist W. S. Burbank spent ten months in the field at Bonanza and produced an epic report, now available as a PDF and well worth a read before heading up here (see Appendix B).

33. Pitkin

Land type: Mountain creeks
County: Gunnison
Elevation: 11,148 feet at Site C
GPS: A - North Quartz Creek: 38.63904, -106.46942
B - Halls Gulch: 38.64891, -106.48184
C - Bon Ton ruins: 38.68281, -106.48134
Best season: Late summer into as cold as you can take it in winter; water gets lowest as temperature drops.
Land manager: Gunnison National Forest
Material: Fine gold, small flakes
Tools: Any
Vehicle: 4WD required for exploring; 2WD OK to Pitkin
Special attractions: Gunnison Pioneer Museum
Accommodations: Pitkin Campground at northeast end of town; Quartz Campground near Site A. Dispersed camping on USFS land; stay clear of private land. Motels and cabins at Pitkin, plus a store and gas. Full services in Gunnison.
Finding the site: From the intersection of Ninth Street and State Street in Pitkin, drive on CR 76 for 0.4 mile; it becomes FR 765. Drive 3 more miles to a giant culvert and then the entrance to Quartz Campground. Just past the entrance to the campground, you'll see good access to Quartz Creek; there was more access before you reached this spot, along with that culvert, but this is a good place to start. To reach Site B, drive farther up the mountain on CR 776/FR 765 for 0.9 mile, and turn left onto FR 766. Drive about 0.8 mile to a good junction where you can access the creek. The coordinates are from the creek crossing. To reach the ruins of the Bon Ton Mine at Site C, resume traveling up the mountain toward Cumberland Pass. Drive about 3.2 miles, and you should have no trouble spotting the mine ruins on your right.

Prospecting

Silver, and then gold, was discovered near Pitkin in 1878. A small crew of prospectors stumbled across a large boulder of rich silver ore at the mouth of Gold Creek, near what is now Ohio City. It was the first major strike on the western slopes of the Rockies. Farther up Quartz Creek, prospectors found additional mineralization, and the boom was on.

The Bon Ton Mine ruins, just below Cumberland Pass on upper Quartz Creek, still contain good ore samples.

Pitkin was originally named Quartzville when it was founded by Frank Curtiss, George P. Chiles, and Wayne Scott in early 1879. It was renamed in honor of Colorado's second governor, Frederick Walker Pitkin, perhaps hoping to sway him for development help. Some specimens in the early days assayed $18,000 per ton. Over thirty mines operated during the winter of 1880. The railroad reached Ohio City in 1881, but the area's mines didn't have staying power, and other than a brief period of excitement in the 1930s, it has been quiet ever since. Koschmann and Bergendahl (1968, p. 101) estimate about 80,000 ounces of gold came from the Gold Brick–Quartz Creek District, which would include mines above Ohio City, the Box Canyon area, and Quartz Creek.

At Site A you can access Quartz Creek easily; several other places along here are open, but this is close to the campground. Bedrock isn't easy to find, but there are plenty of traps and big rocks to dig around. Unfortunately, the creek doesn't go right through camp. At Site B you can access Halls Gulch. The coordinates are on the water; the road goes right, up to Maid Lake and multiple tailings, or you can go left, all the way to the Fairview Mine at 38.66357, -106.53219. The Bon Ton Mine at Site C produced molybdenum,

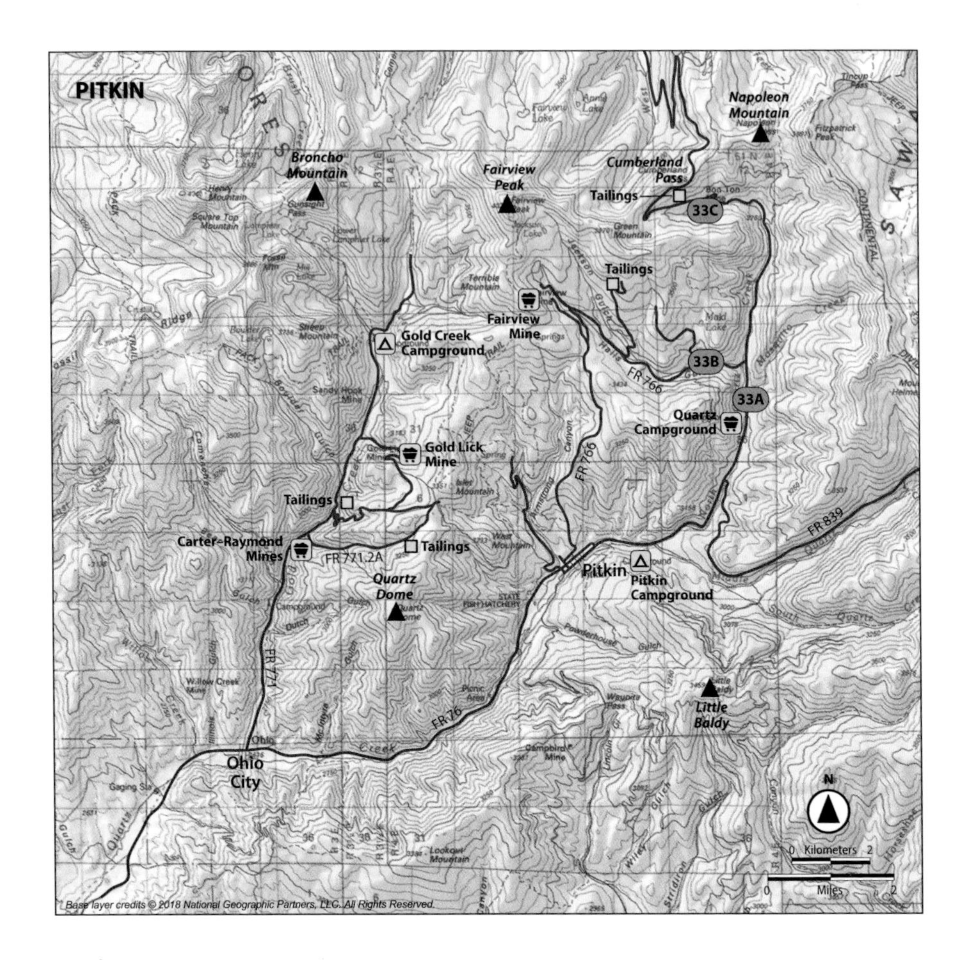

gold, silver, copper, and lead. There are lots of pyrite specimens on the tailings to get you started.

These coordinates barely scratch the surface of this district, but it wasn't a major gold producer, and gold is spotty in the creeks. From Ohio City, you can journey up Gold Creek and check the Sacramento Mine ruins via tough jeep trails at 38.63069, -106.56706 or the easier Carter-Raymond Mines at 38.61228, -106.59625. There is more private land and patented land up there to dodge, so I am only mentioning it here. The headwaters of Tomichi Creek, including Robbins Creek, Bonanza Creek, and Galena Creek, might also be worth checking out.

34. Washington Gulch

Land type: Mountain creeks
County: Gunnison
Elevation: 10,994 feet at Site D
GPS: A - Lower Gulch: 38.93218, -107.01776
B - Upper Gulch: 38.94573, -107.02861
C - Tailings: 38.96545, -107.04068
D - Painter Boy Mine ruins: 38.96854, -107.04133
Best season: Late summer
Land manager: Gunnison National Forest
Material: Fine gold, small flakes
Tools: Pan, sluice
Vehicle: 4WD suggested
Special attractions: Crested Butte hosts numerous festivals for mountain biking, film, wildflowers, and more during the summer. Marble, Colorado, is home to Yule marble, a superb white rock that was used in many buildings in Washington, DC.
Accommodations: Plenty of B&Bs, motels, lodges, and vacation rentals in Crested Butte. Dispersed camping at Sites A and B. Primitive camp on Gothic Road.
Finding the site: From Crested Butte, drive north on Gothic Road for 1.5 miles, then turn left onto Washington Gulch Road. Drive 4.3 miles to a sharp left turn in the road. Look for a junction with roads coming in from the south. Leave the main road and take the best dirt road through a dogleg and then straight to the water, about 0.1 mile, near a clump of trees overlooking the creek. This is Site A. To reach Site B, return to Washington Gulch Road/CR 811 and resume traveling up the mountain. After 0.5 mile a road leads down to the water, where there are usually campers. The next 0.6 mile offers on-again, off-again access, but after a total of 1.1 miles, there is access to the creek again that requires a bit of a hike. To reach Site C, resume traveling up the mountain for another 2.1 miles. These tailings reach the road, and you can find some good minerals here. Just don't go racing up the tailings pile, as this is private property once you leave the road easement—about 10 feet. The Painter Boy Mine ruins at Site D are private, so stay out. There are some great views just another 0.1 mile farther, at the trailhead leading to Paradise Divide.

Prospecting

Miners reached Gothic Mountain above Crested Butte as early as 1874, looking for precious metals. They did well, but silver production waned quickly. The area had the best coal in Colorado, however, and high-quality coal flowed from the area via the railroad by the turn of the century. The coal mines shut down in 1952, leaving ski slopes and summer tourism as the principal revenue sources. An attempt to develop a major molybdenum deposit on nearby Mount Emmons sparked a major battle with local citizens, and it has been on hold since 1977.

You should be able to find good colors in Washington Gulch. The area was placered heavily between 1861 and 1880 and hasn't seen a lot of activity since. There may be enough water to run a sluice; I've read online reports about dredging there, but you'd have a tough time getting enough water, so

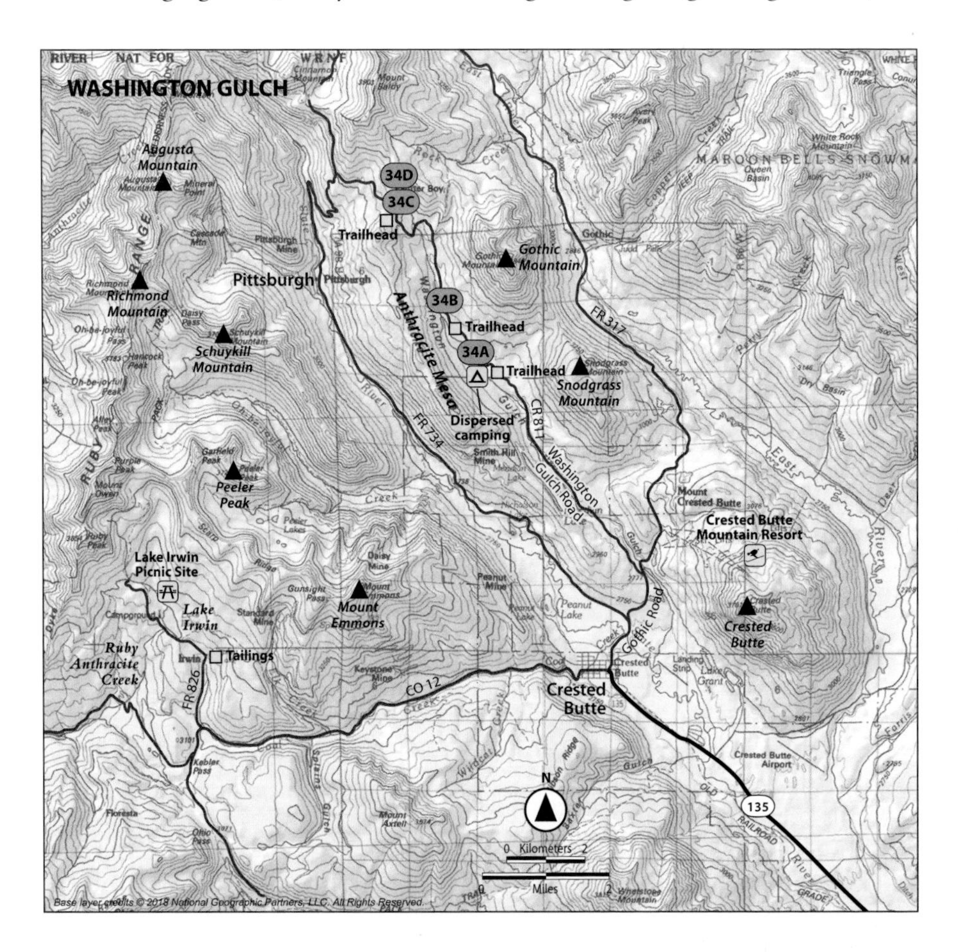

Washington Gulch has good colors and is fairly easy to get to.

stick to hand tools. Site A is a popular camping spot, but you should be able to find a place to the north where the creek comes in. There are plenty of rocks to move around, but bedrock is too far down to reach with a shovel.

At Site B the gulch isn't real close, so you'll have to walk a bit. About a half mile south of the coordinates, there might be decent access, but it's a bit swampy as you go down the hill. A few prospects and minor mines are scattered all through the Site B area, notably in the hills to the west. Soon you'll reach private land, however. As you go up the mountain, you'll reach the town site of Elkton, which was a silver mining camp. It still has one active cabin. The Painter Boy Mine was a very modest gold strike but is on private land and not accessible. The tailings along the road at Site C are associated with the Painter Boy, and there are pyrites in those hand samples.

In the next drainage to the west, on Slate Creek, the old camp at Pittsburg has shown new life, which is too bad, as it means there is a lot of private property to dodge. The Pittsburg Mine at 38.96008, -107.06505 exploited silver, lead, and gold. Finally, at Lake Irwin near Mount Emmons, a town once thrived when the silver ran strong at the Forest Queen, Bullion King, and Ruby Chief Mines. Ruby Anthracite Creek might be worth checking if you're in the area. According to legend, the Lead Chief Mine was discovered by a Missouri farmer whose horse had died. While waiting for a replacement, he hammered away at the nearby rocks, discovering a ledge of lead and silver (Dallas 1985, p. 108).

35. Taylor River

Land type: Riverbank
County: Gunnison
Elevation: 8,842 feet at Site A
GPS: A - Parking area: 38.76202, -106.66212
B - Crash site: 38.76633, -106.65345
C - Lottis Creek: 38.77646, -106.63296
Best season: Midsummer through early fall
Land manager: Gunnison National Forest
Material: Fine gold
Tools: Pan, sluice
Vehicle: Any
Special attractions: Taylor Park Reservoir
Accommodations: Full services in Gunnison. Multiple developed and primitive campgrounds along the Taylor River and at Taylor Park Reservoir.
Finding the site: From downtown Gunnison, drive north on Main Street/CO 135 for 10.1 miles, then turn right onto CR 742. Follow the river up for 14.1 miles to the big pullout across from Lodgepole Campground; this is Site A. To reach Site B, drive another 0.6 mile up the river to a turn to the left. The short gravel loop reconnects to the pavement in a few hundred yards, so if you miss it, just take the second turn. Site C is above the mouth of Lottis Creek, about 1.6 miles farther up the mountain.

Prospecting

The Ute Indians historically made the Taylor River area their stronghold and named it "The Valley of the Gods." Prospectors first entered the drainage in 1860, when Jim Taylor left the Cherry Creek and Clear Creek rushes near Denver and came across Lake Pass to the Gunnison region. According to www.thegeozone.com, Taylor found gold in just about every stream near what is now Tincup.

The Taylor River is no longer regularly replenished due to the reservoir, but there is still color in the gravels, and if you can get deep enough, you should find some larger pieces. Site A is the easiest access, being right along the road at a paved parking area. The river comes through the canyon in more

This marker honors eleven airmen who perished near here in a mysterious training crash during the height of World War II.

or less a straight line, but there are some bigger rocks downstream. It's a good spot for quick sample.

Site B is below the sign discussing the 1943 crash of a B-24 bomber, killing all eleven men on a training mission. Park near the sign, but don't block it. Hike down to the creek on one of the trails, and look for some big rocks. You'll need low water here—the soil near the water is like turf, and while it traps material, it's a pain to break up with your fingers.

Site C is above the mouth of Lottis Creek, and it's always a fun science experiment to sample the creek, then the river, and see which one you like better. Since Lottis Creek is not dammed, it should be better, but the Taylor River drained more gold mines, thus it had more gold. We checked Lottis Creek above the campground near the trailhead, at 38.77194, -106.62203, and panned some good colors. There are other access points for the Taylor River along this stretch, so if you see something worth checking, and you don't see any claim markers, give it a try.

Two well-known "lost mines" are in this area. The slopes of Cross Mountain, near the Lottis Strike of 1861, were said to hold a 10-foot quartz

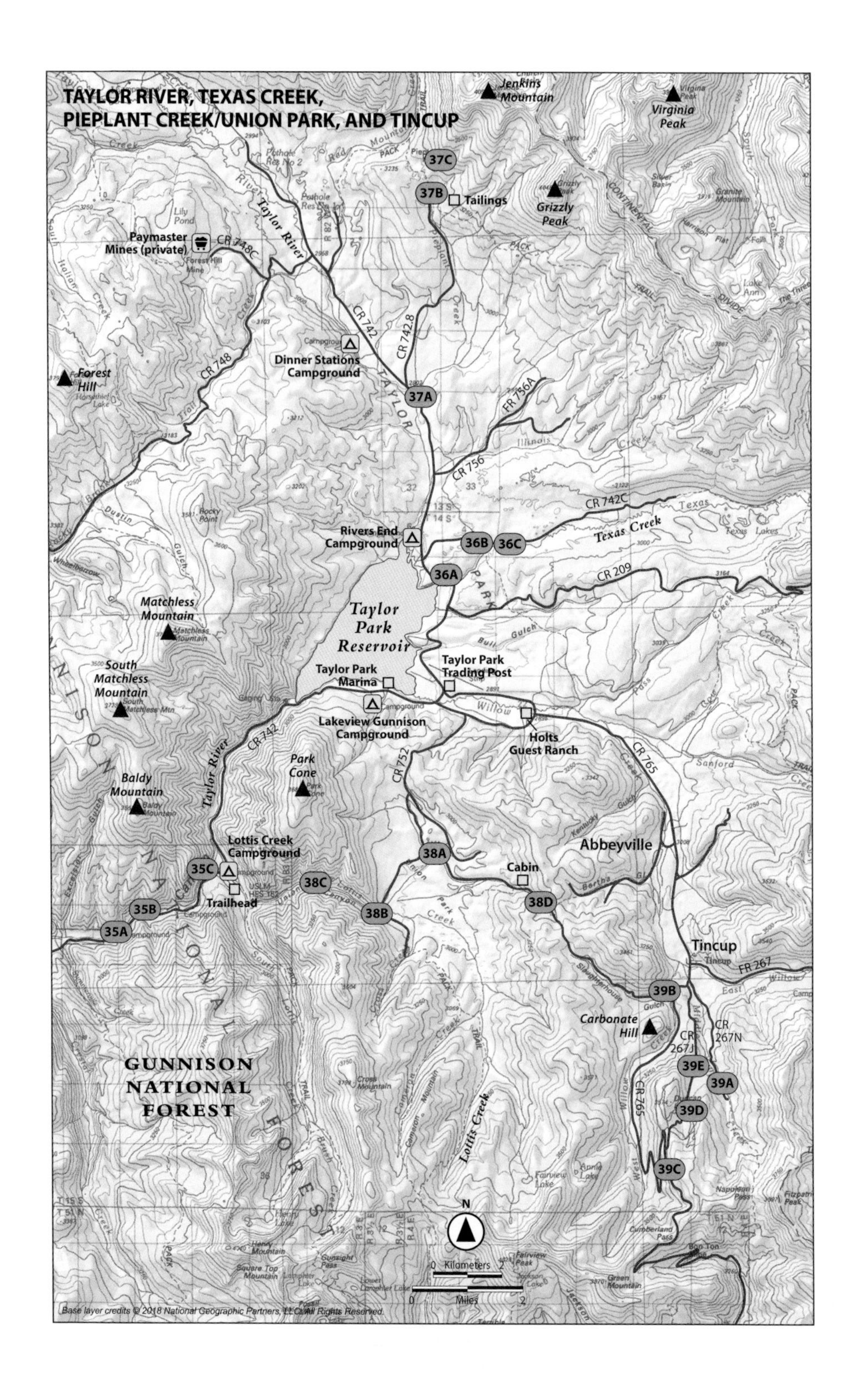
TAYLOR RIVER, TEXAS CREEK, PIEPLANT CREEK/UNION PARK, AND TINCUP
Jenkins Mountain
Virginia Peak
37C
37B
Tailings
Grizzly Peak
Taylor River
Paymaster Mines (private)
CR 748C
CR 742
CR 742.8
Dinner Stations Campground
Forest Hill
CR 748
37A
FR 756A
CR 756
CR 742C
Rivers End Campground
36B
36C
Texas Creek
36A
CR 209
Matchless Mountain
Taylor Park Reservoir
South Matchless Mountain
Taylor Park Marina
Taylor Park Trading Post
Lakeview Gunnison Campground
Holts Guest Ranch
CR 765
CR 742
Taylor River
Park Cone
CR 752
Baldy Mountain
Lottis Creek Campground
38A
Abbeyville
Cabin
35C
38C
Trailhead
38D
35B
38B
35A
Tincup
FR 267
39B
Carbonate Hill
CR 267J
CR 267N
GUNNISON NATIONAL FOREST
39E
39A
CR 765
39D
Lottis Creek
39C
N
Kilometers
Miles
Base layer credits © 2018 National Geographic Partners, LLC. All Rights Reserved.

vein with gold that assayed $440 per ton. In addition, the Lost Murdie Vein, discovered by a Kansas engineer named W. D. Murdie, was spotted in one of the streambeds of Taylor Park. Samples assayed at 2,400 ounces of silver and 95 ounces of gold per ton. When Murdie returned, the stream had resettled its bed and the vein was buried, but it still may be out there.

36. Texas Creek

Land type: Mountain creeks and rivers
County: Gunnison
Elevation: 9,385 feet at Site A
GPS: A - Bottom: 38.84681, -106.55562
B - Ford: 38.85522, -106.54529
C - Camp: 38.85548, -106.53971
Best season: Late summer through early fall; wait for low water.
Land manager: Gunnison National Forest
Material: Fine gold
Tools: Any
Vehicle: Any for Site A. After that, 4WD required; roads are rough and rutted.
Special attractions: Taylor Park Reservoir
Accommodations: Full services in Gunnison. Multiple developed and primitive campgrounds along the Taylor River and at Taylor Park Reservoir. Rivers End Campground is close by. Primitive camping at Site C.
Finding the site: From the intersection of CR 742 and CR 765 near the cafe and ATV rental at Taylor Park Trading Post, drive north on CR 742 for 2.5 miles to a bridge across Texas Creek and easy access; this is Site A. To reach Site B, continue north on CR 742 for 0.7 mile, then turn right onto CO 742C. Drive 1.1 miles, then stay right to continue on CR 742C. After 0.4 mile you will return to Texas Creek. There is a ford here, so access is easy. To reach Site C, drive another 0.3 mile east on CR 742C to the first of many camping areas along the road.

Prospecting

Both Texas Creek and Illinois Creek saw activity in the initial gold rush to Taylor Park in the 1860s and 1870s. The easy gold was quickly exhausted, and there don't appear to be any lode mines above these coordinates. However, it's possible the gold traveled down from the north, as there are lode mines at Pieplant and north Taylor Park. In addition, a cabin and some workings are located up at Magdalene Basin, according to the maps. Here's part of the write-up at TheGeozone.com:

> *"All of these streams have their headwaters in the Sawatch Range which forms the Continental Divide in this part of the state. All of them cut through*

Texas Creek has a heavy runoff through July, so try to time your visit for later in the summer. Don't confuse this spot with the Texas Creek that flows into the Arkansas River near Cotopaxi.

ancient Precambrian granite and metamorphic country rock. Although most of these streams are gold-bearing, the source for this placer gold has never been found. Prospectors may want to focus on the many streams that feed into the Taylor River east and northeast of Taylor Park Reservoir. The area of interest is quite extensive, stretching from Red Mountain Creek southward to Willow Creek. Each of the streams in this part of Taylor Park merits special attention. Prospectors may want to systematically trace the course of each stream from its head to its junction with the Taylor River."

According to Parker (2009, p. 67) the Taylor Park Placers contain "native gold and an assortment of heavy minerals derived from the weathering of granite pegmatites. (These include grains of monazite, zircon, and garnet.) The source for the gold flakes, gold wire, and small nuggets found in these streams has never been found!"

You should also be able to easily pan colors out of Illinois Creek, just to the north. Here are some accessible coordinates if you're in the area: 38.87731, -106.55055.

eplant Creek

Land type: Mountain creek
County: Gunnison
Elevation: 10,850 feet at Site C
GPS: A - Pieplant Creek mouth: 38.88969, -106.56602
B - Pieplant Mill: 38.93876, -106.55943
C - Pieplant Mine: 38.94726, -106.55795 (est.)
Best season: Midsummer through early fall
Land manager: Gunnison National Forest
Material: Fine gold, small flakes
Tools: Any
Vehicle: Any
Special attractions: Tincup
Accommodations: Full services in Gunnison. Developed campgrounds at Taylor Park Reservoir. Dispersed camping throughout USFS lands.
Finding the site: From the intersection of CR 742 and CR 765 near the cafe and ATV rental at Taylor Park Trading Post, drive north on CR 742 for 5.9 miles to a turnout just before the road goes across Pieplant Creek. This will give you easy access to the mouth of the creek and also to the Taylor River. To find the Pieplant Mill at Site B and the Pieplant Mine, resume travel north on CR 742. Drive 0.3 mile and turn right onto CR 742.8. Drive about 3.8 miles to the meadow, and go all the way around it to the ruins and a left turn. It is about 1.1 miles to the mine at Site C, and it might be a lot simpler just to hike it.

Prospecting

Pieplant Creek (sometimes written out as two words) is named for the wild rhubarb that flourishes in the swamps near the old town site. A post office was established here in 1904, around the time that the Woods Mining Company erected a mill capable of processing 200 tons of gold ore per day. At one time there were 100 residents at the mill site in the meadow; a few buildings are still standing, but the area is now used as a cow camp.

The Pieplant Mine may have been a significant contributor to the gold in the area. If you can get deep enough at the mouth of the creek to find larger pieces, they should appear coarse under a hand lens, as they haven't traveled far. There is also a little bit of ore around the old mill site to sample.

The mouth of Pieplant Creek is marked and easy to find; the mine itself is a bit of a hike.

The road follows the creek for a ways before you reach the meadow, but there isn't enough water most of the time to do anything. You could grab some buckets of material if you find a good pinch point or trap, and then work them back at the Taylor River. Close to the mill site, the creek is just a trickle, but much closer to the mine itself, the trail crosses an unnamed creek that might be a good place to bag up some samples and stuff in your backpack. Just for grins, you could sample some of the other tributaries of Pieplant Creek and give them a test, too.

Prospectors James Jenkins and John Lynch are credited with locating the lode deposit here. I didn't attempt the hike up to the mine, but I'm told there are decent ore samples up there, and that it's a steep bit of trail that leads to Jenkins Mountain.

If you want to keep going up the Taylor River, there are several more access points along the way. I tried to get all the way to the Paymaster Mines at 38.92615, -106.63113 and beyond, but ran into a gate and private land. Still, the good news is that there was a pretty significant gold producer at the upper end of Trail Creek, which feeds into the Taylor River, so even the Dinner Station Campground has color in the river gravels.

38. Union Park

Land type: Mountain creeks and rivers
County: Gunnison
Elevation: 9,727 feet at Site A
GPS: A - Nellie Occurrence: 38.77993, -106.55943
B - Lottis Creek crossing: 38.76612, -106.57866
C - Union Canyon: 38.77370, -106.59874
D - Tailings: 38.76881, -106.52671
Best season: Midsummer through early fall
Land manager: Gunnison National Forest
Material: Fine gold
Tools: Pan; sluicing possible
Vehicle: 4WD suggested
Special attractions: Taylor Park Reservoir
Accommodations: Full services in Gunnison; limited services at Tincup. Multiple developed and primitive campgrounds along the Taylor River and at Taylor Park Reservoir.
Finding the site: From the intersection of CR 742 and CR 765 near the cafe and ATV rental at Taylor Park Trading Post, drive south on CR 742 for 0.1 mile, then turn left onto CR 55. Drive 0.6 mile, then turn right onto CR 752. Drive 3.2 miles, go slightly right, and you'll reach Site A in 0.2 mile. To reach Site B, backtrack 0.2 mile, then turn left onto FR 752. Drive 1.5 miles, past the "Cowboy Cabin," and you'll reach the fence and some cliffs. Lottis Creek continues to the west; this spot at Site B offers good access amid several tailings piles. To reach Site C, continue west about 1.3 miles; you'll see the tailings piles from the old workings easily. To reach Site D, backtrack to Site A and continue driving southeast on CR 752 for 2.3 miles, past the placer tailings, a creek crossing, and an old cabin before you reach the tailings at the coordinates.

Prospecting

In October 1859 Jim Taylor panned gold out of Willow Creek, which now empties into Taylor Park Reservoir. Parker (2009, p. 67) reports the first gold discovery at Union Park two years later, in 1861. With so much gold coming out of Willow Creek, the Taylor River, and Tincup, the Union Gulch area was skipped over, but Parker reports that a flume was built around 1898 in Union

Lottis Creek crosses Union Park and carries enough gold to have seen placer operations back in the day.

Canyon, which indicates there was plenty of activity at Site C. Tailings piles can be found all over the place between Site B and Site C, and you should have no trouble pulling colors here. There aren't many boulders, but there is cobble, and the creek twists and bends through the canyon. I was hoping to find more bedrock to scrape, but I didn't. Still, you might be able to get a sluice going, as there is good flow here.

Some reports talk about uranium in this canyon. Don't store your black sands in your front pocket!

At Site A the water can get a bit smelly from the free-range cows roaming the meadow. There is a lot of quartz in the gravels here, and small colors are common, but it wasn't my favorite place to dig. This might be one of those areas that is best to visit when the ground is frozen. Mindat.org shows Site A as the Little Nellie Placers.

Your best bet for bigger pieces is in Union Canyon between Sites B and C. Parker assumes that the source of the gold here is the same mountain range that supplied Willow Creek to the west, such as the head of Bertha Gulch. The topo map for this area shows Gold Creek as a tributary of Lottis Creek, closer to its headwaters. We found a cabin at 38.77263, -106.53277 on the jeep trail that leads to Slaughterhouse Gulch and plenty of placer tailings, and there were lode tailings at Site D. This road is very rough, but it does offer access to the top of Bertha Gulch. That mine appears to be gated, however.

39. Tincup

Land type: Mountain creeks
County: Nevada
Elevation: 10,165 feet at Tincup
GPS: A - New Gold Cup Mine: 38.72454, -106.46955
B - Slaughterhouse Gulch: 38.74773, -106.48803
C - Jimmy Mack Mine: 38.70418, -106.48638
D - Old Gold Cup Mine: 38.71907, -106.47887
E - Tincup Gulch: 38.72927, -106.47885
Best season: Mid to late summer
Land manager: Gunnison National Forest
Material: Fine gold, small flakes
Tools: Pan, sluice; hammer for tailings
Vehicle: 4WD required for Sites D and E
Special attractions: Tincup Boot Hill Cemeteries
Accommodations: Limited services at Tincup; closest full services in Gunnison. Less services at Pitkin, over Cumberland Pass. Developed campgrounds at Taylor Park Reservoir and Mirror Lake. Dispersed camping OK if away from private land.
Finding the site: Tincup is most easily reached via Gunnison, but if the pass is open, you can get there from St. Elmo. From Gunnison, drive north on CO 135 for 9.9 miles, then turn right onto CR 742. Drive 23 miles, then turn right onto CR 765, near the Taylor Park Trading Post. Drive another 1.9 miles, then stay left to remain on CR 765 and follow Middle Willow Creek to Tincup. To reach the Gold Cup Mine at Site A, drive south on Grand Avenue/CR 765 for 0.3 mile, then turn left onto CR 765N. Stay on this road for about 2.1 miles, past the cemeteries, to reach the ruins. The rest of the coordinates are to the west; there is a faint 4WD track leading over the ridge to them, but the more "civilized" route is to backtrack to CR 765 and turn left. Drive 0.5 mile on the main road, then turn right onto CR 765K. Drive about 0.3 mile, park, and walk in to the tailings on the right (north). To reach the Jimmy Mack Mine at Site C, return to CR 765, the main road, and continue up the mountain following West Willow Creek toward Cumberland Pass. At 3.4 miles the switchbacks start. There is a small road to the ruins of what is likely the Blistered Horn Mine at 38.70275, -106.49198; you'll see the tailings along the main road after 0.8 mile. In another 1.9 miles, you'll see a mine shaft on the left side of the road, then after

another 0.3 mile is the turn for CR 765F. To reach the top of Tincup Gulch, stay on CR 765F for 0.8 mile, turn right onto CR 765H, and drive another 0.8 mile, then turn left onto CR 765J for 0.1 mile to the ruins. Finally, to reach Site E, head back toward Tincup. Drive 0.3 mile to the intersection with CR 765G, and take a sharp right to stay on CR 765J. Drive another 0.8 mile, past the pond, and look for the creek. The coordinates are just a short walk from water—if there is any, as this is a seasonal creek. An easier way to reach Tincup Gulch would be to just take CR 765J up the gulch for 2.3 miles from south of Tincup; a harder way would be to drive all the way to Cumberland Pass and take CR 765J from there.

Prospecting

James Taylor followed the gold up the Middle Fork of Willow Creek to Tincup Gulch in 1860 (Clark 2009, p. 67). He was said to have used a tin cup to hold his gold, hence the name. Taylor and his party returned in 1861. Three other prospectors—two Dutch brothers named Karl and Fred Seigul and their companion, Fred Lottis—followed Taylor's party out of Granite, hoping to get in on the early stages of a gold rush, but lost their trail. They found good strikes on their own, however, north of present-day Ohio City, at what became Dutch Flats and the Gold Brick District. Lottis is credited with locating the Union Park District, and named the creek that runs through there.

Parker (2009, pp. 66-67) reports that the most important placer workings came from what topo maps show as the Old Gold Cup Mine: "The placers of Tincup Gulch are all in the upper section of the gulch, above Middle Willow Creek valley. They are found from place to place in the lower half mile of this upper section; in this distance the workings were almost continuous the last 1,000 feet. The gravels were shallow and all placers were worked to bedrock. The gold was derived from the important Gold Cup ore deposit, which crops out on the east slope of the Gulch, and some small veins on the slopes above it."

You can try panning West Willow Creek at Site A or Slaughterhouse Gulch at Site B, but Tincup Gulch is the best bet for panning up here. Even the gravels at Site E are rough and pointy—so is the gold. This area was mostly worked by hand before miners gave up in the 1880s. Park at the coordinates for Site E and walk to the creek. Follow it downhill and look for any place you can get to bedrock. The other coordinates will give you a taste of the many mines in the area.

Tincup started under the name Virginia City, but miners insisted the Tincup name reflected the town's true heritage, and it stuck. Early life there

These tailings at Slaughterhouse Gulch are your best bet if you don't have a sturdy 4WD to reach the upper mines.

was rough; of the first eight sheriffs, one completed his contract safely, one quit, one was fired, one went nuts, one became a preacher, and the rest were shot dead. During the long winters, supplies had to come in via skis from St. Elmo, about 13 miles away. In 1882 around 3,000 people lived there, but by 1884 the boom was over.

A dredge operated on Willow Creek between Bertha Gulch and Kentucky Gulch, and the tailings are easy to spot. There is so much private land out here that all the plans I had for Bluebird Lane, Fisherman Drive, and Bertha Gulch were quickly dashed.

One final note: Hubnerite, one of the telluride varieties, is occasionally found on the dumps in this district.

40. Chalk Creek

Land type: Riverbank
County: Chafee
Elevation: 11,184 feet at Site E
GPS: A - Chalk Creek trailhead: 38.71657, -106.19952
B - Nice bend: 38.71527, -106.25148
C - Turn for Mount Antero: 38.70998, -106.29168
D - St. Elmo: 38.70349, -106.34524
E - Mary Murphy Mill: 38.66729, -106.35789
Best season: Late summer
Land manager: San Isabel National Forest
Material: Fine gold, small flakes
Tools: Pan, sluice
Vehicle: 4WD recommended to Site D and required to reach Site E
Special attractions: Prospecting supplies at the Rock Doc Rock Shop, CR 260 and US 285 (38.67696, -106.09299)
Accommodations: Full services in Salida and Buena Vista. Lots of developed campgrounds along Chalk Creek.
Finding the site: From just south of the little town of Nathrop, about 17 miles north of Salida on US 285, drive west on Chalk Creek Drive/CR 162 for 6.8 miles to Bunny Lane and turn left. Follow the road for just 0.1 mile and you'll see the Colorado Trail parking area; just past there is another trailhead parking area that leads to the creek below the bridge. This is Site A. Site B is a nice bend in the creek bed about 3.2 miles farther up the mountain on Chalk Creek Drive. Look for a trail to the creek. Site C is just the coordinates for the turn up to the world-class aquamarines at Mount Antero, included here for convenience. Just prior to this turn is the right turn for CR 292, which gives you access to the north side of Chalk Creek, both east and west. To reach St. Elmo, drive up the mountain on Chalk Creek Drive another 3.4 miles. The road is gated at the west end of town, but you should at least check out St. Elmo if you've never been there. To reach Site E, backtrack 0.3 mile and take a sharp right onto CR 295. Drive up for 2.9 miles, then turn left onto CR 297. The ruins are 0.4 mile up this road.

Prospecting

The gold in Chalk Creek is primarily from the Mary Murphy Mine. Koschmann and Bergendahl (1968, p. 92) credit the Mary Murphy Mine with continuous operation from 1870 to 1925, producing 220,000 ounces of gold, worth $4.4 million then (or about $266 million today), plus considerable silver, lead, and zinc. Probably 75 percent of the district's production came from the Mary Murphy. Older guidebooks (Johnson, 1971) show the most interesting placer ground in the lower part of the drainage, but that's mostly private land and gated off. The hot springs resorts control what's next, and then you start getting into more public land. The access point at Site A should provide some colors, and it's shady, too.

Multiple campgrounds in the area provide creek access, such as at Chalk Lake. Above there, Site B offers good access to the creek, and there weren't any claim markers in 2017. The first bridge for CR 292 just above here offers some good access to the creek.

I included the coordinates for the turn to Mount Antero as a courtesy to the rockhounds who want to check it out. If you've watched *Prospectors* on the Weather Channel, you know all about the Busse family, the Cardwell family, Amanda Adkins, and Steve Brancato. Several good rockhounding guidebooks are available for more information about the gemstones found here; *Rockhounding Colorado* and *Colorado Rockhounding* are both listed in appendix B.

Gem hunting aside, keep going up the mountain to St. Elmo. Even if you're in a hurry to reach the Mary Murphy area, this thriving "ghost town" is worth the side trip. In the summer it's crawling with tourists, ATVs, Jeeps, and even mountain bikes, but that just adds to the charm.

The Mary Murphy Mill is photogenic and a good place to rest after the jolting drive up the slow, bumpy road. There are good specimens in the road, but I found better ones at the ruins to the south a tenth of a mile, over in the trees. The source for the ore at the mill came from the mine at 38.66824, -106.35113. The Iron Chest Mine is even farther up the mountain, at 38.67175, -106.34883. You can pan the creek below the mill and get colors, or swing a hammer at the many tailings piles.

If you had kept going on FR 295 to the end of the road, you'd have reached the Hancock Placers near the site of Hancock, at 38.64046, -106.36121. The Allie Belle Mine is along this road, at 38.64976, -106.36759.

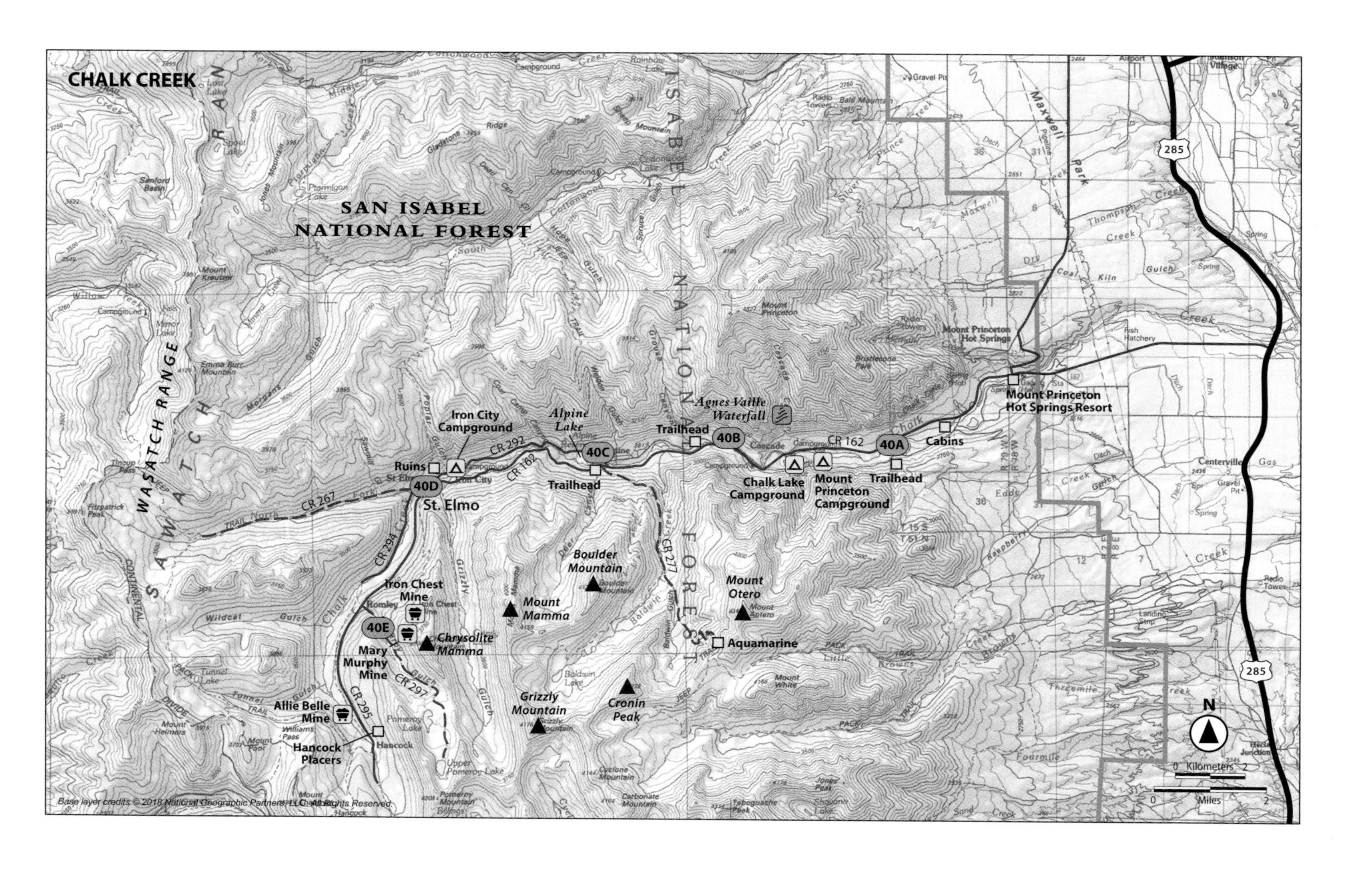
CHALK CREEK
SAN ISABEL
NATIONAL FOREST
SAWATCH RANGE
WASATCH RANGE
285
Mount Princeton Hot Springs
Mount Princeton Hot Springs Resort
Agnes Vaille Waterfall
Trailhead
40A
40B
40C
40D
40E
Cabins
CR 162
CR 292
CR 267
CR 277
CR 294
CR 295
CR 297
Chalk Lake Campground
Mount Princeton Campground
Alpine Lake
Iron City Campground
Ruins
St. Elmo
Boulder Mountain
Mount Otero
Aquamarine
Mount Mamma
Iron Chest Mine
Chrysolite Mamma
Mary Murphy Mine
Grizzly Mountain
Cronin Peak
Allie Belle Mine
Hancock Placers
N
Kilometers
Miles
Base layer credits © 2018 National Geographic Partners, LLC. All Rights Reserved.

Ruins of the Mary Murphy Mill, above St. Elmo. It's a rough drive to get here, but completely worth it. The mine itself is even farther up the mountain.

There is a good report on this district by Dings and Robinson (1957, pp. 83-85) USGS Professional Paper 289. They report that pyritic quartz veins up to 50 feet thick cross Chrysolite Mountain. Galena, sphalerite, pyrite, chalcopyrite, calcite, rhodonite, rhodochrosite, barite, and fluorite occur locally. The best ores were strongly oxidized and are typically brown, porous limonite or limonitic quartz with variable amounts of cerussite, calamine, and smithsonite and patches or grains of galena. Free gold reportedly occurs in much of the ore.

41. Buena Vista

Land type: Riverbank
County: Chafee
Elevation: 8,270 feet at Railroad Bridge
GPS: A - Miners Camp: 38.88439, -106.15259
B - Railroad Bridge Campground: 38.92321, -106.16989
Best season: Late summer, or later when water is lowest
Land manager: BLM and San Isabel National Forest
Material: Fine gold, small flakes
Tools: Pan, sluice
Vehicle: 4WD recommended for Site B
Special attractions: Buena Vista Heritage Museum (Court and Main Streets)
Accommodations: Full services in Buena Vista. RV parking, developed campgrounds, and primitive campsites all along the Arkansas River.
Finding the site: From the intersection of Main Street/CO 306 and US 24, head east on E. Main Street for 0.2 mile and turn left onto N. Colorado Avenue. Drive 0.3 mile and continue onto CR 371. Take CR 371 for 3 miles to the campground entrance. This is Site A; river access is under the railroad and down to the water. Look closely to see if any new claim markers have been put up; it was open in 2017, but things change quickly. To reach the Railroad Bridge Campground, drive another 2.9 miles north on CR 371. The campground is easy to spot, and is full on holidays. To reach this spot from the north, drive south from Granite on US 24 for 7.5 miles, then turn left onto CR 371 and drive 3.4 miles to Site B.

Prospecting

Henderson (1926, p. 9) credits prospectors with uncovering placer gold in the Arkansas drainage as early as 1859, primarily on Cache Creek and Clear Creek. Koschmann and Bergendahl (1968, p. 91) report that prospectors by 1860 had found paying gravels from Buena Vista southeast for 25 miles to the Fremont County line, as well as north of Granite into Lake County. In all, miners were working Lost Canyon Gulch, Chalk Creek, Cottonwood Creek, Pine Creek, Bertscheys Gulch, Gold Run Gulch, Gilson Gulch, Oregon Gulch, and Rietchey's Patch. Through 1969 about $400,000 was recovered by placer mining. By 1923 Henderson credits miners in Chafee County with

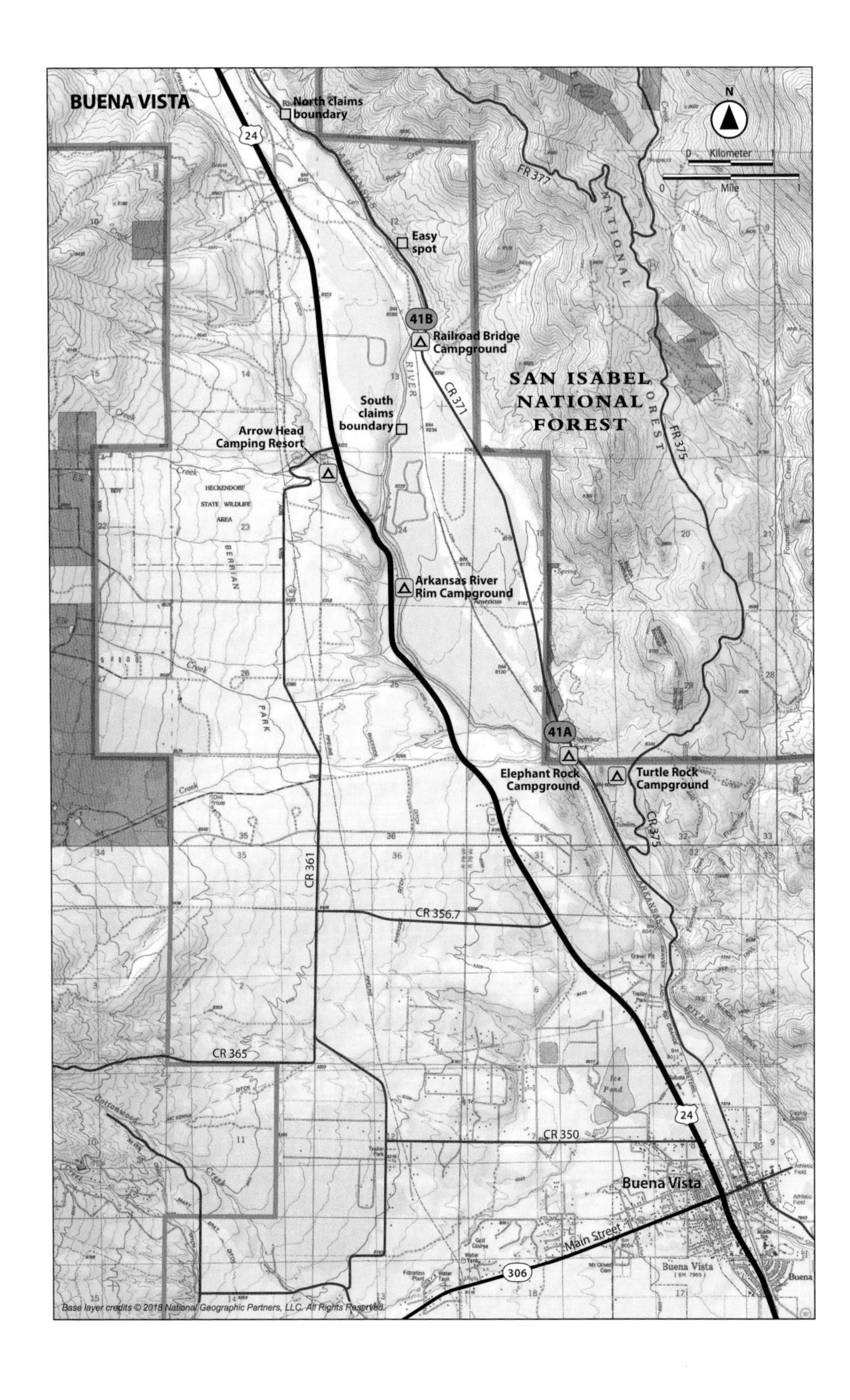
BUENA VISTA
North claims boundary
Easy spot
41B
Railroad Bridge Campground
SAN ISABEL NATIONAL FOREST
FR 377
FR 375
CR 371
South claims boundary
Arrow Head Camping Resort
HECKENDORF STATE WILDLIFE AREA
Arkansas River Rim Campground
41A
Elephant Rock Campground
Turtle Rock Campground
CR 375
CR 361
CR 356.7
CR 365
CR 350
24
Buena Vista
Main Street
306
N
Kilometer
Mile
Base layer credits © 2018 National Geographic Partners, LLC. All Rights Reserved.

Railroad Bridge is a popular spot—it's been in the GPAA claims guide for many years.

recovering about $7.4 million in gold, with $1.5 million of that by placer miners.

Because the Arkansas River is so popular, it is claimed for many stretches along its reach from above Granite to below Buena Vista. Some of the campgrounds along this stretch are actually open, because claim owners don't want the hassle of chasing people off all the time, but you should always check for signs. Since claim markers are also sometimes vandalized by unscrupulous miners (not readers of this book, I hope!), you can't always rely on the absence of signs either. You can check with the BLM or US Forest Service local offices—they are very helpful. But for now, try these two sites.

Site A is easy to find, and while there was a claim marker south of the camp along the road, it indicated that it marked the northwest corner of that claim, so this area is open. Site B is a Gold Prospectors Association of America (GPAA) claim known as the Railroad Bridge, but listed as the Arkansas Group in their claims guide. If you already belong to GPAA, dig away. If you don't, and have never visited one of their club claims before, go ahead and give it a try for a few minutes and check a few pans. If you're camping at the locale anyway, so much the better. If you're just cruising through, skip the camping area and go about 0.25 mile north to 38.92629, -106.16913. This is a good day-use spot, although I've seen a tent in the parking area. You can try a little farther up at a slightly better primitive camping spot right along the road at 38.93122, -106.17244.

There are 440 acres in the Arkansas Group of claims, stretching from below the Railroad Bridge at 38.91382, -106.17221 all the way north to 38.94391, -106.18641. One way to recon the claims is to just slowly drive up and down, noting areas where there are big boulders, bedrock, bends in the river, parking, camping, etc. During holidays this entire stretch of river fills up fast, and many of the people have no idea there is good gold panning here—they just want to raft or play in the river. Other times, the only people out here are club members.

My favorite spot is a quarter mile north of the Railroad Bridge. Even when the water is high, you should get colors in every pan. When the water is low, there is still plenty of flow to set up a sluice. Then you can dig away under the big rocks and bring buckets of pay dirt to the sluice. Dredging is not allowed here; the rafters need to be able to pass through, and they seem to have priority.

42. Clear Creek

Land type: Mountain creek banks, outdoor museums, tailings
County: Chafee
Elevation: 10,253 feet at Site G
GPS: A - Clear Creek Reservoir Campground: 39.01846, -106.27771
B - Savage Placers: 39.00252, -106.33937
C - Primitive camps: 39.00041, -106.36246
D - Vicksburg Outdoor Mining Museum: 38.99837, -106.37582
E - Crescent Mining Camp: 38.99176, -106.41147
F - Upper Camp: 38.98873, -106.41793
G - Winfield: 38.98499, -106.44118
H - Banker Mine: 38.96623, -106.45771
Best season: Late summer
Land manager: San Isabel National Forest
Material: Fine gold, small flakes
Tools: Pan, sluice
Vehicle: 4WD recommended above Site A
Special attractions: Winfield
Accommodations: Full services in Buena Vista and Leadville. Developed campgrounds and primitive camping areas along Clear Creek.
Finding the site: From Granite on US 24, drive south for 1.7 miles, then turn right onto CR 390. If coming from the south, this turn is about 15 miles north of Buena Vista on US 24. Drive 1.7 miles to the campground on the left and try to park close to the water. This would make a good base camp for exploring the entire Clear Creek drainage. To reach Site B, return to CR 390 and drive up the mountain for 3.7 miles. You should see dirt tracks leading to the cliffs above the water, or walk down to the west where the terrain is less rambunctious. The primitive camps at Site C are about 1.3 miles up the road, past the Dawson Cabin. Site D, the Vicksburg Outdoor Mining Museum, is another 0.9 mile up. Next is Site E, the Crescent Mining Camp, about 2.1 miles farther up the mountain. Just 0.4 mile up is Site F, a large primitive camping area above the water. The ghost town of Winfield is about 1.3 miles farther, and we hadn't seen any claim markers for a long time when we got to this point. Winfield is Site G. Take the left fork and follow the main stem of Clear Creek all the way to Site H, the Banker Mine, on about 1.7 miles of rough road.

Prospecting

Clear Creek had plenty of gold at the bottom of the drainage where it enters the Arkansas River, but the best ground is now flooded by Clear Creek Reservoir. There is a small stretch from the dam to the river, with good parking for the anglers. We found numerous claim markers there, as it's an obvious spot, being right on the river where a rich creek enters.

Site A appears to be open; there were no claim markers, and one person was panning the creek. The coordinates are in one good spot, but there are good areas everywhere here.

Site B is the site of the Savage Placers, and it too appears to be open. One advantage here is the amount of bending the creek does along this stretch, along with the large boulders. Site C and Site F are similar, with primitive camping, good access, and plenty of colors. The photography spots along this route are easy to access, at the Dawson Cabin, the Vicksburg Outdoor Mining Museum, the Crescent Mining Camp, and Winfield.

According to legend (see www.ghosttowns.com), Vicksburg (Site D) was founded after prospectors from Leadville tried their luck along the creek.

Beautiful ore sample from the tailings pile at the Banker Mine.

View of the ruins at the Banker Mine from atop the tailings pile. There are plenty of ore samples here.

Their burros wandered off to get water from the mountain stream, and when the prospectors found them, they also checked the gravels, which showed promise. The mining museum here is definitely worth checking out; while you're there, make a small contribution to keep it going. The Crescent Mining Camp (Site E) is photogenic, and you can access the creek below here where the road fords the creek.

Winfield (Site G) was known as Florence, then Lucknow, then Winfield. The "wandering burros" story about Vicksburg may actually apply to Winfield. Not much gold was found here; copper and silver were the main commodities, up until 1918. Now the town is maintained by the Clear Creek Canyon Historical Society, and it makes for a leisurely stroll to wander around the buildings and take photographs.

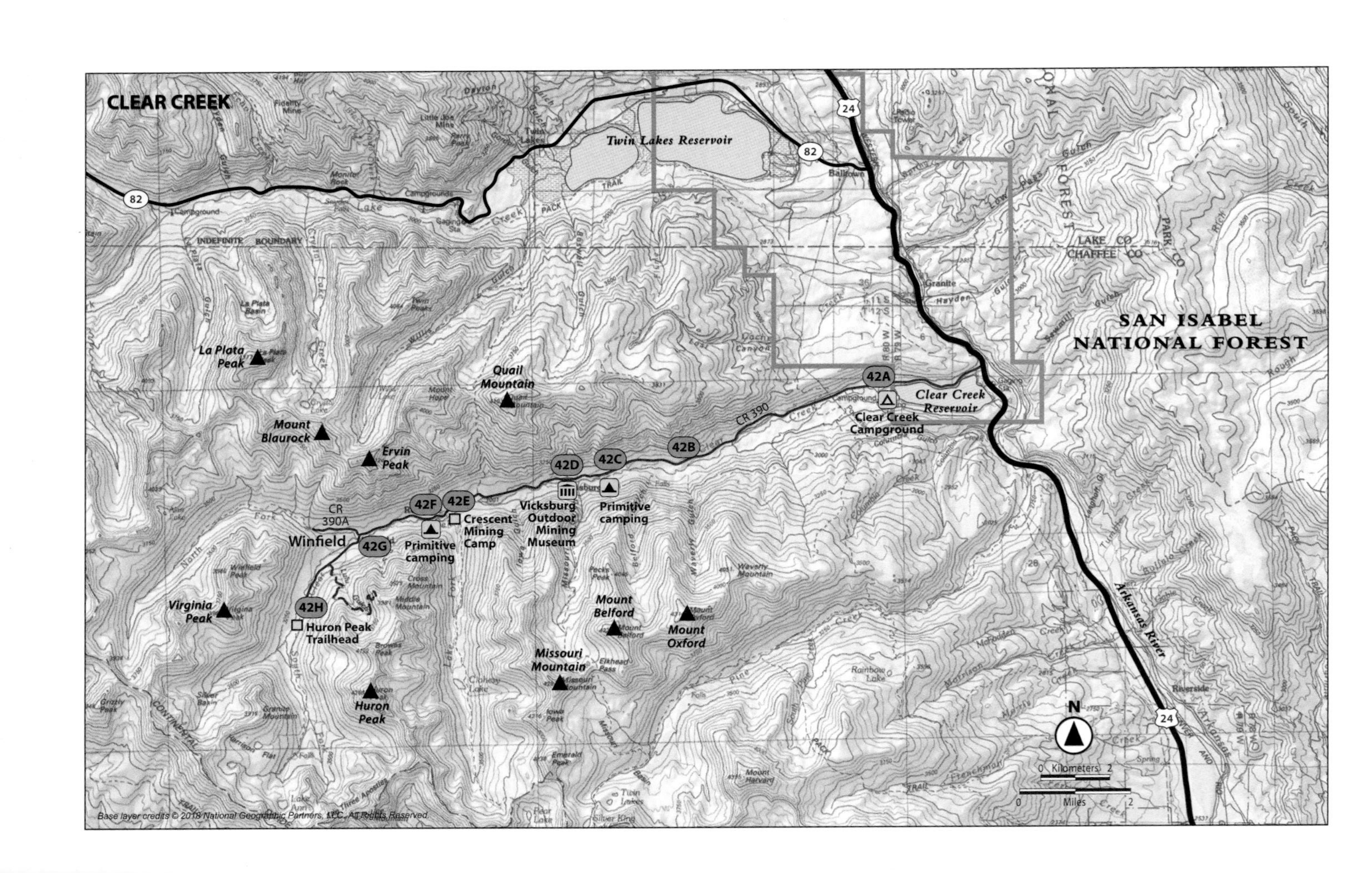
CLEAR CREEK
Twin Lakes Reservoir
82
24
SAN ISABEL
NATIONAL FOREST
LAKE CO
CHAFFEE CO
PARK CO
La Plata Peak
Quail Mountain
Mount Blaurock
Ervin Peak
42A
42B
42C
42D
42E
42F
42G
42H
CR 390
CR 390A
Clear Creek Reservoir
Clear Creek Campground
Vicksburg Outdoor Mining Museum
Primitive camping
Crescent Mining Camp
Primitive camping
Winfield
Virginia Peak
Huron Peak Trailhead
Huron Peak
Mount Belford
Mount Oxford
Missouri Mountain
Arkansas River
N
Kilometers
Miles
Base layer credits © 2018 National Geographic Partners, LLC. All Rights Reserved.

The Banker Mine (Site H) was located by Fred Aude and Godfried King; two of Aude's children are buried at Winfield. The Banker Tunnel took almost ten years to complete and reached 3,700 feet long. Ore containing silver and lead was shipped from 1910 to 1927. After that, the machinery was looted or moved; only the bunkhouse and office buildings remain. By 1917, when geologists for the state of Colorado visited to work on *Molybdenum Deposits of Colorado*, the adits and tunnel were flooded and the air was already dangerously bad. Just by inspecting the dumps, they found bismuth ore, fluorite, beryl, galena, molybdenite, muscovite, pyrite, quartz, and rhodochrosite. Here are some clues about what rocks to look for, from their report:

> *The gangue is pure white massive quartz. Scattered irregularly through it, but chiefly in layers near the walls, is molybdenite, most of which occurs as fine, dark lead gray grains, and aggregates sheet-like in form. The richest streaks are on the vein walls and in the country rock near the walls, where with silvery mica and pyrite there is some very high-grade ore. Considerable bismuthinite was found on the dump in a white quartz exactly like that which contains the molybdenite, and in one instance bismuthinite and molybdenite were found together (Worcester 1919, pp. 41–42).*

43. Cache Creek Public Area

Land type: Creek bank
County: Chafee
Elevation: 9,302 feet
GPS: 39.03717, -106.30202
Best season: Spring through late summer
Land manager: BLM
Material: Fine gold, small flakes
Tools: Pan, sluice
Vehicle: Any, with care
Special attractions: Granite
Accommodations: Full services at Leadville and Buena Vista. Primitive camping at site.
Finding the site: From Granite on US 24, there is a dangerous turn behind the rafting headquarters on the west side of the road onto Lost Canyon Road/CR 398. This stretch is very narrow; I had to back up for traffic coming down twice in three days. The road is full of washboards in stretches, too. In all, from US 24 in Granite go 2.1 miles on Lost Canyon Road/CR 398 to the power lines; your GPS might tell you to turn earlier, but don't. Turn left at the power lines, rather than proceeding straight on CR 398. Follow this dirt track for 1.2 miles and hope for a good camping spot.

Prospecting

Cache Creek's gold deposits were formed during the last ice age as a gravel outwash terrace from the lower Bull Lake glacial period. One probable source is the Lost Canyon Placers (Parker 2009, p. 59). Parker describes deposits in Sections 4 and 5, Township 12 South, Range 80 W, with a level of pay gravel on bedrock ranging up to 15 feet thick, with gold often still clinging to quartz and very rough in aspect. Note that they are held in a patented claim.

Prospecting began at Cache Creek in 1858; Campbell and Shoewalter are credited with the first excavations and pits. By 1863 a ditch was completed, and for the next two years about 200 people lived in several camps during the most productive times. In a 2016 BLM proposal on converting the Cache Creek site to a fee-dig area, the report indicates there were 4,000 feet of Long Toms and sluices here in 1867, with the workings extending 150 feet wide

Beaver pond near the bottom of the Cache Creek public area. The creek is running to the left of this photo.

and 30 feet deep. Water from Lake Creek was diverted to the camp, so that by 1884 a complete system of tunnels and flumes provided water to six hydraulic monitors. In 1911 Cañon City and Pueblo sued over environmental damage from sedimentation, and mining shut down.

Cache Creek is your best bet for a public panning locale in Colorado. It's open, easy to find, covers more than 2,100 acres, and has plenty of gold, plus you're sure to meet some interesting fellow prospectors who can help you out. Everyone we met here was extremely helpful. There was a little grumbling about high-banking being banned after being allowed for so long, but even a sluice and bucket system can recover good gold here.

You can find numerous YouTube videos from Cache Creek prospectors setting up sluices and running pay dirt. Some swear by the blue layer, while others find the reddish-brown layer pays better. One prospector with a sizable sample jar confided that he believes there are numerous "false bedrock" traps between layers, but he finds the gold intermittent, so he digs around.

The site has limited facilities: one Porta-Potty, decent fire rings, and plenty of parking and camping. The most coveted camping spots are on the trails that lead to the lower reaches of the creek. One camper reported a bear sighting while we were there for a week in July, so use caution with your food. There are no garbage services, so be prepared to haul out your trash.

The future will be interesting here; the BLM has already imposed many restrictions on the use of pumps and motors, marked off trees to avoid digging

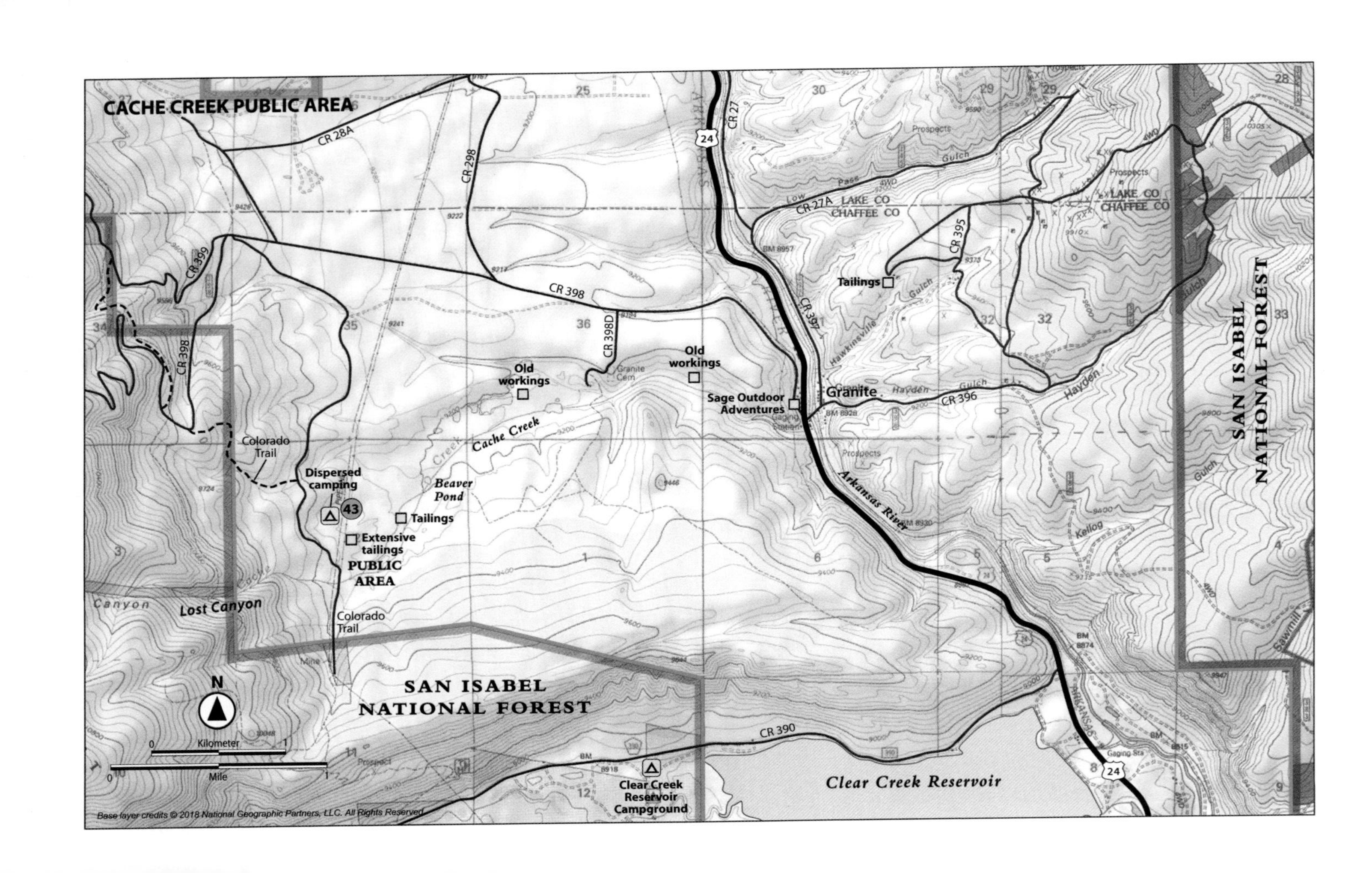

Base layer credits © 2018 National Geographic Partners, LLC. All Rights Reserved

Cache Creek is Colorado's premier public panning location. These chunks came from about three hours of work.

under them, banned handcarts and wheelbarrows, and set up barriers to prevent driving closer to where you want to work. It's true that mining is messy and there are holes everywhere. Some of that damage pre-dates the public area, as the area was heavily hydraulicked and the scars are everywhere—huge piles and fields of large boulders dot the landscape. Without millions of dollars spent bulldozing, landscaping, and remediating, it will always be obvious that this is the site of an old placer mine. By concentrating mining activity here, and giving people a place to go where they are free from worries about mining claims, land-use restrictions, and private property, you could argue that this is a far better and much cheaper model. Still, you can probably expect permitting and fees in the future.

The tailings piles contain a lot of good-looking quartz with rusted staining, and it might be interesting to use a metal detector up here after breaking apart some of the quartz. There are so many rusty nails and other chunks of iron in the lower area, you might be better off running a detector higher up the drainage.

44. Lake Creek

Land type: Creek bank
County: Lake
Elevation: 9,276 feet at Site A; 12,205 feet at Site B
GPS: A - Lower access: 39.06439, -106.40188
B - North Fork: 39.08191, -106.53938
C - South Fork: 39.05501, -106.51168
D - Stewart Mine: 39.04185, -106.57771 (est.)
Best season: June through Sept
Land manager: San Isabel National Forest
Material: Fine gold, small flakes
Tools: Pan, sluice
Vehicle: 4WD required for Sites C and D
Special attractions: Independence Pass
Accommodations: Full services in Leadville. Developed campgrounds around the lake and up the creek. Dispersed camping options galore.
Finding the site: From Leadville, drive south on US 24 about 14.3 miles and turn right onto CO 82W. If coming from Granite, drive about 2.6 miles north on US 24 and turn left. Drive west on CO 82 about 8.3 miles and look for easy parking on the right. To reach Site B, continue up the mountain for 8.2 miles, then turn left onto CR 82D. Drive about 0.1 mile to plenty of primitive camp spots along the creek, below the concrete structure. To reach Sites C and D, backtrack down the mountain for 2 miles, then turn left onto CR 82C. Drive about 0.9 mile, then continue on CR 399. Follow CR 399 about 0.2 mile to the Site C coordinates. There is easy access to the creek here. To continue to Site D, drive another 2.2 miles up the gulch, past the turn to the right for Sayres Gulch, and turn left on FR 394 to go another 2.5 miles. The last part is probably gated.

Prospecting

Twin Lakes was once the go-to resort town for the upper crust of Leadville, featuring palatial homes and cozy cottages, and a resort and limited services are still available there. I didn't have much luck finding information about the three mining areas along Lake Creek—probably because they never amounted to much. Just past Twin Lakes there is a right turn at 39.08183, -106.38609

Lake Creek runs fast and clear above its confluence with the South Fork.

that goes steeply up the mountain, swings left to run along Gordon Gulch, swings left even harder, and turns into a hiking trail to get over to the Little Joe Mine (39.08815, -106.40894) at the top of Gordon Gulch and the White Star Mine (39.08157, -106.40949) below Parry Peak. Here's part of the report on the Gordon District from Cappa and Bartos (2007, p. 41):

> *There are at least eight veins in the area . . . all of which shipped gold and silver ore to a mill in Twin Lakes by way of a gravity tram. Early production records are not available. Incomplete records from the American Smelting and Refinery Company smelter at Leadville from 1937 to 1942 indicate that 163 tons of crude ore averaging 1.6 oz per ton gold, 4.8 oz per ton silver, 9.7 percent lead, 2.1 percent zinc, and 0.7 percent copper were shipped from*

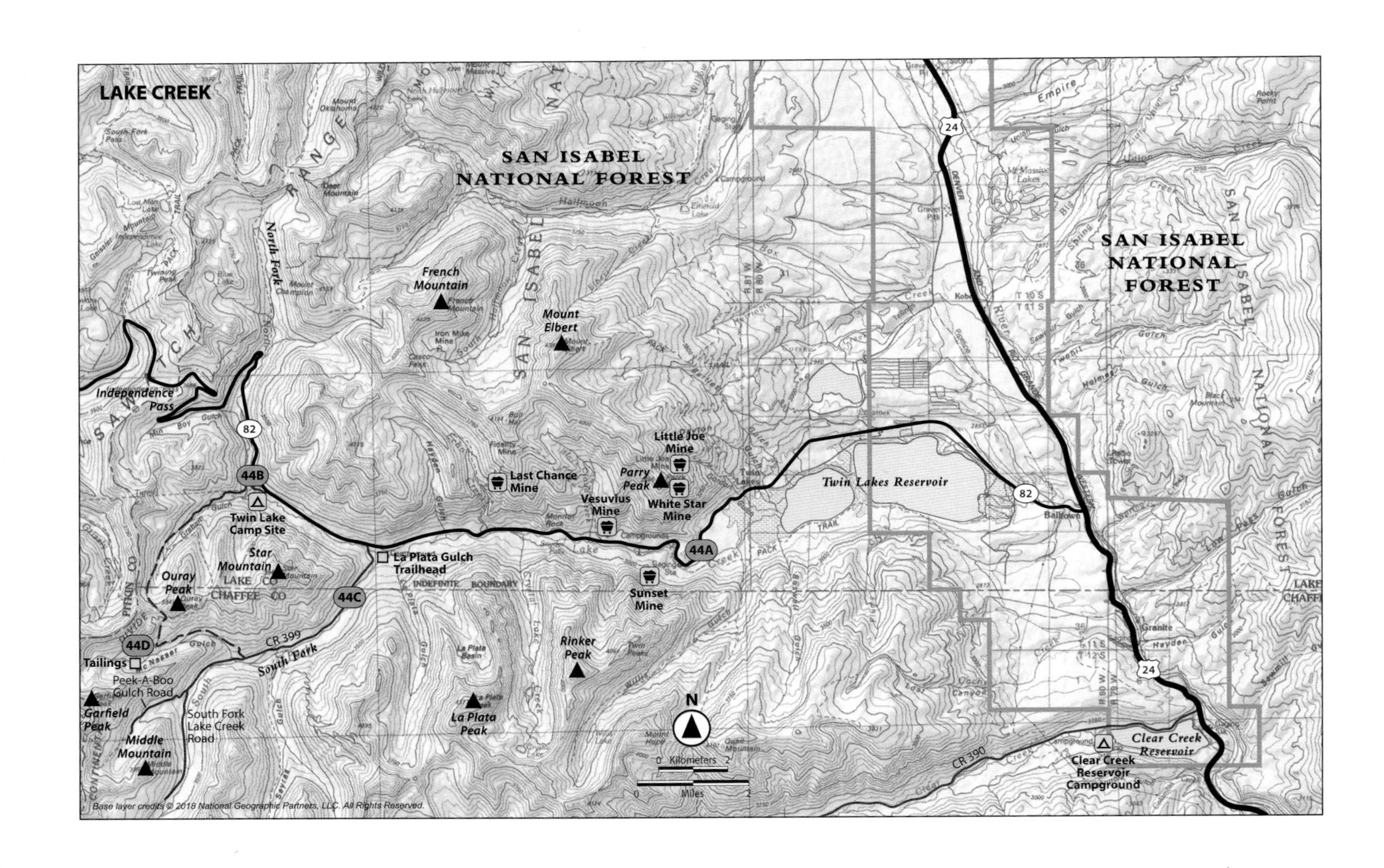
LAKE CREEK
SAN ISABEL NATIONAL FOREST
SAN ISABEL NATIONAL FOREST
French Mountain
Mount Elbert
Independence Pass
82
44B
Twin Lake Camp Site
Last Chance Mine
Little Joe Mine
Parry Peak
Vesuvius Mine
White Star Mine
Twin Lakes Reservoir
82
Star Mountain
La Plata Gulch Trailhead
44A
Ouray Peak
44C
Sunset Mine
44D
CR 399
Tailings
Peek-A-Boo Gulch Road
Rinker Peak
La Plata Peak
Garfield Peak
South Fork Lake Creek Road
Middle Mountain
N
0 Kilometers 2
0 Miles 2
CR 390
24
24
Clear Creek Reservoir
Clear Creek Reservoir Campground
North Fork
South Fork
Base layer credits © 2018 National Geographic Partners, LLC. All Rights Reserved.

the Gordon Mine. Production at the Gordon Mine was apparently stopped by Federal order during World War II; there is no evidence that it reopened after the war.

There is a lot of patented land here, though it might make an interesting day hike. So that was strike one. As for strike two, I had high hopes for the Vesuvius Mine, at 39.07174, -106.43117, but that was private, gated, and off-limits. And then for strike three, the Sunset Mine, halfway up Sunset Gulch, was gated. Its coordinates are approximately 39.05938, -106.41701. Echo Canyon has the Fidelity Mine, Golden Fleece Mine, and Last Chance Mine; a trail leads up Echo Canyon, and the first mine is the Last Chance, at 39.08361, -106.46589.

All of these locales were minor gold occurrences, contributing to the placers of lower Lake Creek. The two lower campgrounds, Parry Peak and Twin Lakes, offer access to the creek, or you can use Site A. I may have missed an interesting spot at 39.07799, -106.29863 that looked like good access from the maps.

There is a little color at Site B. Mineralization from a few scattered mines, and perhaps even the south side of the Champion Mine, feeds into the North Fork.

Site C is along the South Fork of Lake Creek and leads to the Red Mountain District. The water here is tainted yellow from mine waste and some natural discoloring. This is not a very attractive spot to grab a sample pan, but there was a lot of activity up here that I wanted to check out. Alas, this was as far as I got. Peekaboo Gulch, on the South Fork of Lake Creek, has many mines that contained gold, with by-products of zinc, lead, and silver. Sayres Gulch also has mines, and discolored water. I ran out of daylight and didn't get to explore beyond Site C. The road is rough in most places and very slow going, but there are numerous mining relics, shacks, and tailings piles to investigate; many different drainages that all lead into the South Fork; and very, very few prospectors make it up this far. So that's why I've included it.

Writing for the US Geological Survey in Professional Paper 138, Charles W. Henderson had two stories that are worth sharing. About Lake Creek he wrote:

In 1861 a company built a dam on Lake Creek and flumed the Arkansas River at Georgia Bar for 1,000 feet at a cost of $10,000. They were nearly ready to commence sluicing the bed of the stream, which is very rich, when

> *their dam gave way and carried off their flume. The men immediately enlisted in the First Colorado Volunteers, and no attempt has since been made to flume the Arkansas, although the sand for a hundred miles from its source, taken up on a shovel and panned down, gives a fine color. (Henderson 1926, p. 131)*

He filed this report on the Red Mountain area:

> *Red Mountain seems to be in a belt of lodes, some 3 miles in width, which here crosses the range in the true course northeast and southwest. From the top of the Red Mountain at the head of the left fork of Lake Creek, other red mountains can be seen both to the east and west. Eight miles west, in an air line, a Boston company did some work in 1866, finding similar ore to that found here . . . In the streams and where the creek escapes from the mountains, numerous well-defined lodes have been discovered, not greatly different in width, lineal extent, and character of ores from those of other parts of the Territory. Like them, too, they vary in richness. Some of them are absolutely barren, and some contain $100 gold to the ton . . . (Henderson, 1926, p. 131)*

Nearby Cottonwood Creek did not pan out, as they say. You can try a few spots along the road or drive up to the top, but there just wasn't much color wherever I tried.

45. Independence

Land type: Mountain creek, valley
County: Pitkin
Elevation: 10,965 feet at Site D
GPS: A - Lost Man Campground: 39.12156, -106.62449
B - Independence Historic District: 39.10854, -106.61339
C - Independence Mine: 39.10845, -106.61266
D - Upper Independence: 39.10679, -106.60355
Best season: June through Sept
Land manager: White River National Forest
Material: Fine gold, tailings
Tools: Pan, sluice
Vehicle: Any
Special attractions: Aspen; Holden/Marolt Mining and Ranching Museum in Aspen; Ashcroft Ghost Town
Accommodations: Full services in Aspen. Developed campgrounds up the Roaring Fork River, such as Site A, and around Aspen. Dispersed camping on USFS land.
Finding the site: From Aspen, drive southeast on Cooper Avenue/CO 82 for about 13.4 miles toward Independence Pass. (It's paved all the way over the pass, so any vehicle is fine.) Look for the campground entrance on the right; this is Site A. To reach Site B, drive farther up the mountain on CO 82 for 1.4 miles. You'll see the historical markers on the right. This is the beginning of the Independence Historic District. Site C is across the highway from Site B, and Site D is about 0.5 miles east from Site B.

Prospecting

Mines on West Aspen Mountain had an estimated production of 25,000 ounces, minor compared to nearby Leadville but still important in its day. Independence was the first settlement established in the Roaring Fork Valley, after gold was struck in the vicinity on Independence Day 1879. Technically, these miners were forbidden by Governor Pitkin from mining in the area west of the Continental Divide. The Ute Indians jealously protected their lands, but the lure of gold caused prospectors from Leadville to fan out in all directions.

Looking down the valley in the Independence Historic District. The tailings are on the other side of the road, and you can't pan in the creek until you get below the historic district and private land.

Reports on the actual discovery of the Independence District vary: According to the filing with the National Register of Historic Places, "some say that it was Billy Belden who was leading a group of prospectors to the Aspen area, others claim that it was Charles Bennett, while still other sources point to the peripatetic Dick Irwin."

The first cabin in the camp is thought to have been erected by J. B. Connor. A small tent city sprang up after the discovery of the Geld Placer on the Roaring Fork, and the citizens quickly held a miner's meeting to organize the district. By 1880 about 150 people lived in the camp, and wooden structures started going up. In 1881 the Farwell Consolidated Mining Company out of Leadville bought up the best claims and erected a stamp mill. Farwell

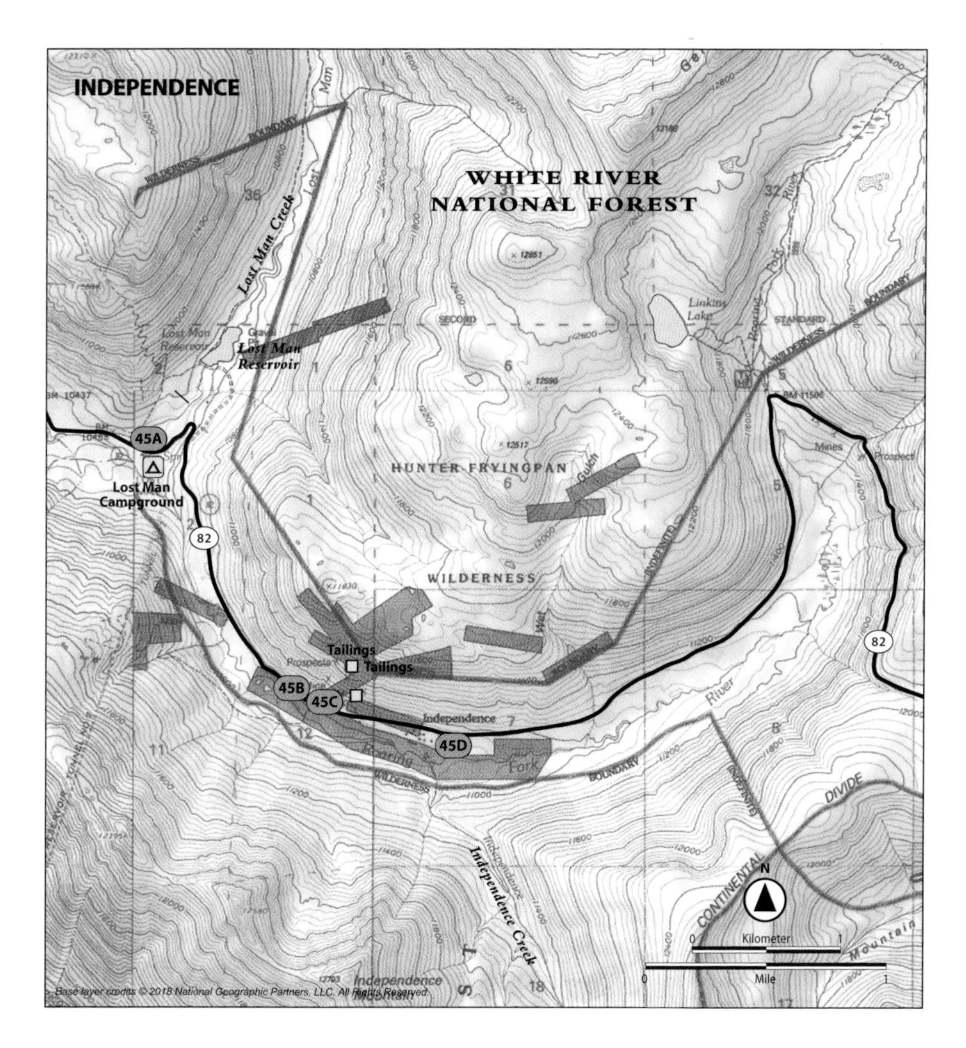

already had two excellent mines in Leadville, so he rolled the dice on a third. Thanks to his investment the town prospered, with seven restaurants, three saloons, and a local newspaper. The mines produced about $190,000 in gold and silver, mostly in the first years, but when Aspen overtook Independence in size and importance, the small town's elevation, harsh winters, avalanche threats, and declining ore deposits spelled doom. After 1912 it was abandoned completely.

In 1973 this area was recognized as a historic district and listed on the National Register of Historic Places as the Independence and Independence Mill Site. The log cabins are now on land partially in White River National

Forest. Over the past twenty years, the remaining buildings were restored and the interpretive kiosks were put in place. The portion of Independence closer to the river is on private land owned by an entity referred to as the Loughren Trust—an organization with no apparent web presence. It's very difficult to find a good map of the historic district, too. I listed Site A as the nearest campground on the Roaring Fork, and it's a good base to scout around for a sample. Your best bet is above the containment structure, which you can find if you walk the creek upstream. I collected some good pyrite samples at Site C, above the highway; there are many more gray and yellow tailings piles up the steep hillside. I listed the east end (Site D) and west end (Site B) of the historic district for reference; you should not pan in the district itself, but downstream a ways from Site B is OK.

Nearby, Aspen is one of Colorado's gems, and it has a lot to offer in the summer. Aspen was the leading silver producer in the world at one time, with the fabulous Smuggler Vein, on the east end of town, supplying one single silver boulder weighing almost a ton. But its stubborn ores gave the owners fits. The Holden/Marolt Mining and Ranching Museum preserved much of the Holden Lixiviation Mill, built in 1891. Lixiviation is a tongue twister of a word that means "to separate a substance into soluble and insoluble constituents by the percolation of a liquid." The mill crushed, heated, and salted the ores, but the experimental process never quite proved out. The 1893 Silver Panic put an end to the mill.

46. Halfmoon Creek

Land type: Mountain stream
County: Lake
Elevation: 10,578 feet at major creek crossing
GPS: A - Easy test: 39.15023, -106.45201
B - Garnets: 39.15152, -106.45783
C - Ford: 39.15371, -106.46365
D - Champion Mill: 39.13706, -106.50439 (est.)
Best season: Late summer is best for these sites; you'll need low water and no snow to reach Site D.
Land manager: San Isabel National Forest
Material: Fine gold, small flakes; loads of garnets
Tools: Pan, sluice
Vehicle: 4WD recommended; road is slow and rough. 4WD with very high clearance and a snorkel required for Site D.
Special attractions: Leadville
Accommodations: Full services in Leadville. Developed and dispersed camping all the way up Champion Creek to the ford.
Finding the site: From downtown Leadville, drive southwest on US 24 about 4.1 miles, to the big bend in the highway before it straightens out. Turn right to go west on CO 300 and drive about 0.8 mile, then turn left onto Halfmoon Road/ CR 11. After 1.3 miles turn right to stay on Halfmoon Road, which soon turns into CO-110 after the river gauge. Drive 7.3 miles up the mountain, on increasingly bad roads. Site A is easy to find, just past several primitive camps and right on the water. To reach Site B, continue up about 0.3 mile. This spot is loaded with garnets, and you can see some excellent samples of garnet gneiss along the road. To reach the ford at Site C, continue up the road about 0.4 mile. There is primitive camping here and good access to the creek. To reach Site D at the Champion Mill, one of the most photogenic of all remaining Colorado mine structures, cross that ford and drive about 2.8 miles. The road doesn't dead-end at the mill—you can keep going all the way to the flanks of Mount Champion, another 2.5 miles, at about 12,800 feet.

Prospecting

There are decent colors and interesting garnets in Halfmoon Creek, but the big attraction by far is the Champion Mill, one of the most photogenic of the remaining mine structures in Colorado. It is remote, difficult to access, majestic, and challenging. This is one of those examples where four-wheeling and gold panning intersect—right at Site C. If you have the wrong vehicle, do not try to ford the creek. If you have the right vehicle and you make the attempt at the right time of the year, no problem!

I watched the USGS water gauge at https://waterdata.usgs.gov/co/nwis/rt go down all through the late spring and early summer, and I thought July would be a good month to attempt it. But there were thunderstorms all month, and since I was alone, with no winch, I chickened out. As I was panning garnets at Site B, a parade of four-wheelers came up the mountain in various Mad Max contraptions, and we chatted for a bit. One thing for sure they had going for them was the confidence of youth. I'm sure they made it.

The Gold Rush Expeditions website (www.goldrushexpeditions.com) lists a 20-acre property for sale (only $22,000) just east of the Champion Mine, on Lackawanna Gulch, called the Mount Massive Gold Mine and Camp. They list two reasons why this area is largely unexplored: short summers and that ford. Here's how they describe it:

> *There is no other way up to the basin than through the river. This river in mid-low runoff is 24–36 inches deep and nearly 25 feet across as you need to navigate slightly up the river to access the other bank. In heavy runoff the water is an estimated 5' deep and roughly 30' across. It is also very fast moving. Experienced off road drivers will have no trouble crossing in full size vehicles equipped with 4WD and a high mounted snorkel. UTVs, ATVs and motorcycles will need to exercise caution at this crossing and only do so when they feel the conditions are safe. As a result of this river crossing, the only traffic the site sees is mostly hikers and historians hoping to see the Champion Mill before it collapses.*

The Colorado Geological Survey described the Mount Champion Mine recently(Cappa and Bartos (2007, p. 41):

> *The property was discovered in 1881, but was not developed until 1907 when the Mount Champion Mining Company purchased the property and started construction of a 50-ton/day mill and a 6,100-ft-long tramline. Significant production started in 1912 and continued until 1918. U.S. Bureau of Mines*

The ford at Halfmoon Creek leading to Site D will stop most drivers in their tracks as they study the route. Just hook a left around that big tree—no problem!

records indicate that 4,759 tons of direct shipping ore were produced at a grade of 3.23 oz. per ton gold and 2.6 oz. per ton silver. In addition, 40,259 tons of milling ore assaying 0.374 oz. per ton gold and 0.28 oz. per ton silver were mined. This ore also had minor amounts of lead (1 percent) and, locally, copper (0.3 to 1 percent). During this period, the Mount Champion Mine generated 26,500 oz. of gold and a total estimated metal value of $550,000 to $600,000. The Mount Champion Mine was dormant from 1919 to 1936. From 1937 to 1940, the property was leased, and there was small-scale mining; incomplete records suggest that 17.5 tons averaging 2 oz. per ton gold and 2 oz. per ton silver were mined (G.L. Fairchild, unpublished report, 1974). There is no record of mining activity past 1941. In the late

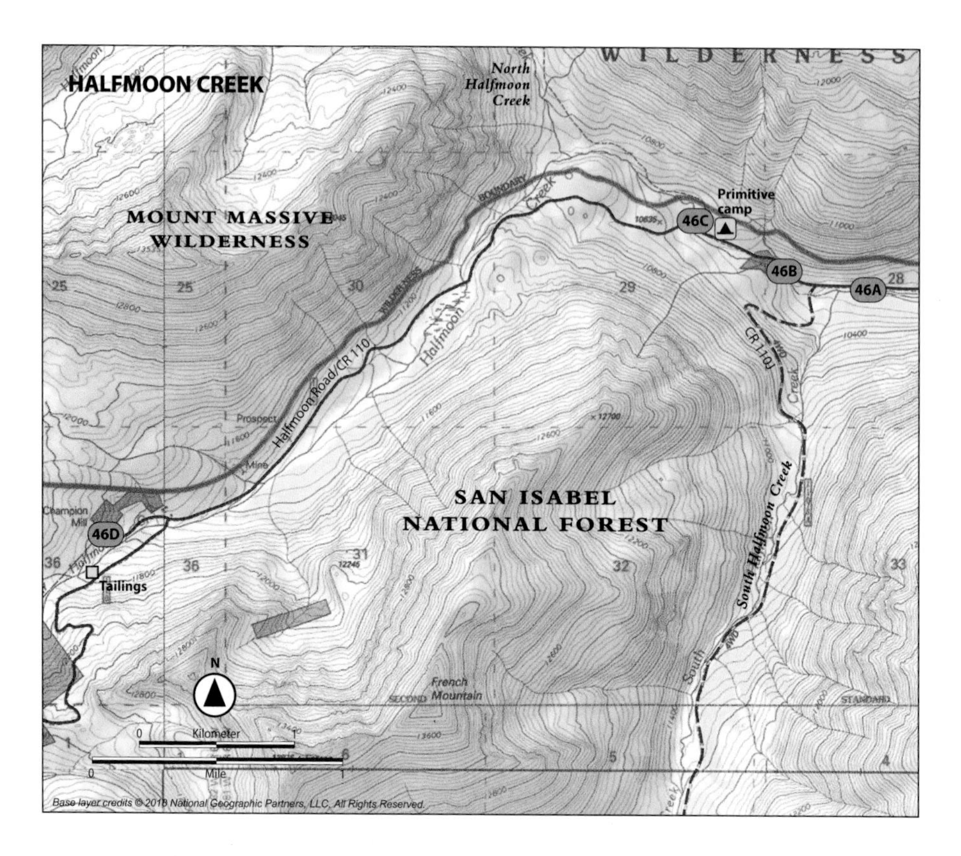

1970s through the early 1980s, the property was further explored with five drill holes, totaling 400 ft.

Halfmoon Creek flows from the flanks of Mount Champion, and it has good colors all along its upper reach. While in the area, check out the abandoned ghost town of Derryall. Massive tailings are visible at 39.12711, -106.32978, and the buildings come into view about 0.7 mile southwest of the coordinates, on CR 24. This land is private and gated, so you'll need a powerful lens to bring the buildings into focus. Note that as of 2017, the property was for sale, so maybe a local real estate agent can give you a tour.

47. National Mining Hall of Fame and Museum

Land type: City museum
County: Lake
Elevation: 10,191 feet
GPS: 39.25096, -106.29411
Best season: Open daily during summer 9 a.m. to 5 p.m.; closed major holidays; closed on Mondays from November through April.
Land manager: City of Leadville
Material: Displays
Tools: Camera
Vehicle: Any
Special attractions: Historic downtown Leadville; Locomotive 641 at Eighth and Hazel (39.25089, -106.28721)
Accommodations: Full services in Leadville, including hotels, motels, B&Bs, hostels, and RV parks. Developed campgrounds nearby, such as those at Turquoise Lake.
Finding the site: From downtown Leadville, drive north on US 24/Harrison Avenue to the big right turn that leaves downtown. Turn left, instead of following the highway right, at E. Ninth Street. The Mining Hall of Fame is just a block west; multiple signs help you find it.

Prospecting

The National Mining Hall of Fame and Museum at Leadville (www.mining halloffame.org) is a must-stop for gold panners. It was founded in 1977 and moved to the former school building in 1987, where it grew to be recognized as "The Smithsonian of the Rockies." The museum has world-class exhibits of gold nuggets, crystalline gold, and the many ores of gold, silver, lead, zinc, and copper. There are other beautiful specimens from Colorado, such as rhodochrosite, a superb gem of deep red from the Sweethome Mine in Alma, along with mineral displays from around the world. The museum contains the Prospector's Cave, a replica of a lode mine, with dark "tunnels" and cramped quarters with old and new mining equipment.

Entrance to the National Mining Hall of Fame.

The Hank Gentsch Diorama Room contains a large series of dioramas depicting the discovery and development of the Clear Creek Canyon mines at Idaho Springs by George Andrew Jackson and John H. Gregory. Jackson prospected so early in the year that he had to build a huge bonfire on a gravel bar where Chicago Creek enters Clear Creek just to thaw the ground enough to pan. While his weary dogs Drum and Kit watched, he cleared off the embers and recovered a large nugget and an ounce of fine gold. In his diary, he recorded his efforts: "Feel good tonight, dogs don't." When Gregory and his team discovered rich gold deposits on Clear Creek, he was said to exclaim, "My wife will be a lady and my children will be educated!"

Elsewhere, there are exhibits on historic European mining techniques, coal mining, space mining, molybdenum uses, Colorado's famed Yule marble, and more. My favorite was the Gold Rush Room, with loads of maps, gold exhibits, photographs, and an enclosed panning station that will show you

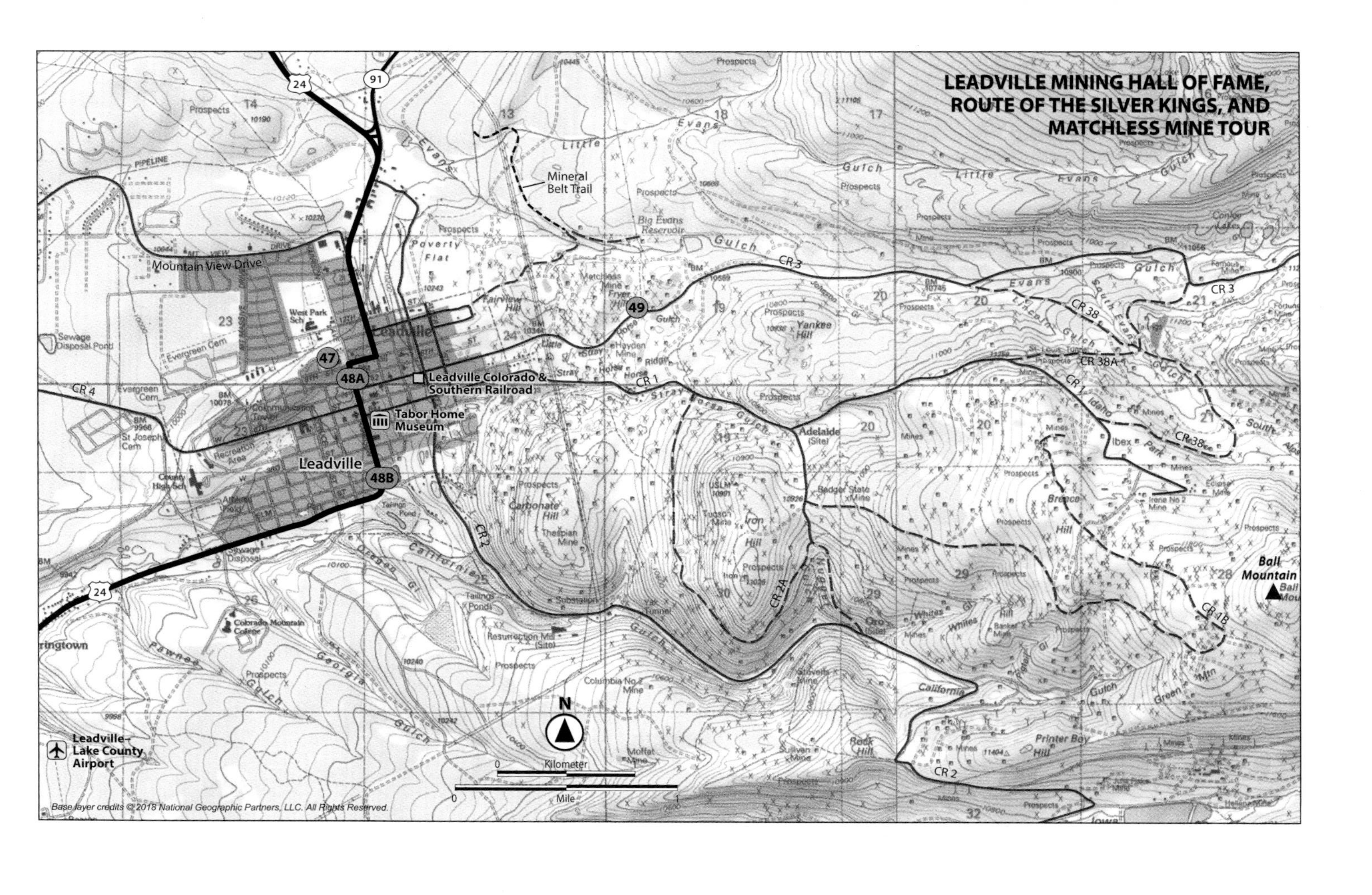
LEADVILLE MINING HALL OF FAME, ROUTE OF THE SILVER KINGS, AND MATCHLESS MINE TOUR
Mineral Belt Trail
Mountain View Drive
Leadville
Leadville Colorado & Southern Railroad
Tabor Home Museum
47
48A
48B
49
24
91
CR 4
CR 3
CR 1
CR 2
CR 2A
CR 38
CR 38A
CR 1B
Adelaide (Site)
Ball Mountain
Leadville-Lake County Airport
N
Kilometer
Mile
0
1
Base layer credits © 2018 National Geographic Partners, LLC. All Rights Reserved.

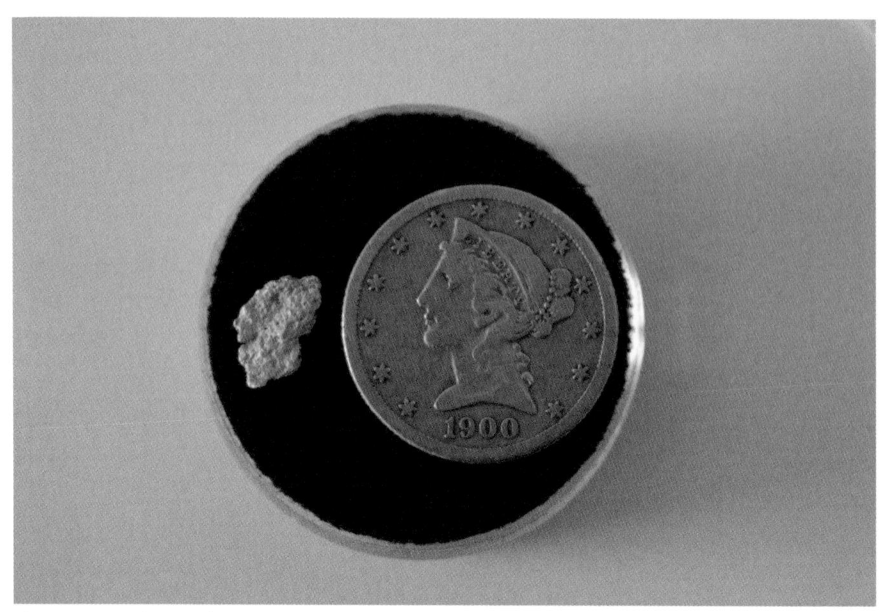

This 0.36-gram beauty from Park County cost $32 at the museum gift shop. Shown next to a 1900-S five-dollar gold piece for scale.

how easy it is to pan gold in a small plastic tub with big rubber gloves—as long as there is gold in the pay dirt. Sadly, you don't get to keep what you can pan out.

Upstairs is a collection of portraits of the men and women who have been elected into the Hall of Fame. An excellent gift shop is downstairs, with lots of books plus a wide assortment of gold nuggets for sale. All in all, it's a fantastic way to spend time with your family or friends.

48. Route of the Silver Kings

Land type: Historic Mining District near Leadville
County: Lake
Elevation: 10,152 feet at Leadville city center
GPS: A - Leadville Lake County Visitor Center: 39.25059, -106.29248 (809 Harrison Ave.)
B - Starting point: 39.24375, -106.29009 (Monroe and Harrison)
Best season: Summer
Land manager: Mostly private
Material: Old mines, mills, cabins, and tailings
Tools: Camera; no panning or rockhounding (private property)
Vehicle: Any; 4WD recommended for the last stops
Special attractions: Leadville
Accommodations: Full services in Leadville, including hotels, motels, B&Bs, hostels, and RV parks. Developed campgrounds nearby, such as those at Turquoise Lake.
Finding the site: The Leadville Lake County Visitor Center is in downtown Leadville on Harrison Avenue. It's easy to find—look for the Golden Burro Cafe and walk a little farther up Harrison. The starting point at the intersection of Harrison and Monroe is at the south end of Harrison where US 24 makes a big bend to go straight through town. Look for the Mountain Peaks Motel.

Prospecting

Leadville was one of the largest silver camps in the world in its heyday. The combined value of gold, silver, lead, zinc, copper, and molybdenum produced from the Leadville mines is estimated to be over $5 billion in current US dollars. Koschmann and Bergendahl (1968, p. 104) describe the early discovery, credited in some accounts to a prospector named Abe Lee:

> *At the time of the "Pikes Peak excitement," some of the early prospectors, searching for gold in stream gravels, wandered across the Rampart, Tarryall, and Mosquito Ranges into South Park and the Arkansas River valley. Early in the spring of 1860, placers were discovered in Iowa and California Gulches, tributaries of the Arkansas River, in what was to become the Leadville district. News of the rich discoveries spread with amazing rapidity, and by July 1860*

the placer camp called Oro City boasted a population of 10,000. The placers, though rich, were quickly depleted, and within 3 or 4 years only a few hundred of the more persistent souls remained. It was reported that $2 million in gold was taken out the first summer.

Once the placer gold started to dwindle, attention turned to lode gold, but it soon petered out as well. Miners soon learned that the heavy gray mineral that was clogging their placer operations was cerussite, $PbCO^3$, or lead carbonate. With a specific gravity of 6.5, cerussite can fill the riffles and allow gold to escape a sluice box. In the case of Leadville's cerussite, silver atoms occasionally took the place of lead in the crystal lattice, making the mineral even more valuable, if harder to extract. In 1868 the first rich silver lodes revealed themselves, and Leadville mining began in earnest. Miners located more silver mines in the early 1870s on Iron Hill and Carbonate Hill, east of town. In 1878 prospectors discovered an extremely rich silver lode at the Little Pittsburg Mine on Fryer Hill, and by 1880 the population of Leadville had jumped to 40,000.

In 1999 the American Smelting and Refining Company (ASARCO) closed the Black Cloud Mine, the last operating mine. There have been, and still are, multiple environmental projects north of Leadville to fix some of the problems the miners left behind. Some of those have reclaimed the best placer opportunities, unfortunately; in addition, there is very little public land to list in this guide. About the nearest spot I found was at East Tennessee Gulch, and it wasn't that great. You might be able to find some open area way up Iowa Gulch on CR 2B. So pack up a picnic lunch and use the Route of the Silver Kings to take a self-guided tour through the area.

The best place to start is Site A, the Leadville Lake County Visitor Center, where you can pick up a map with explanatory text. Then drive back down Harrison Avenue to Site B and enjoy the tour. The chamber of commerce asked me not to include GPS coordinates for all the stops, as they would prefer that visitors stop by the visitor center and get updated information first, so I am honoring their request. I'm also supposed to remind you that rockhounding and panning are not allowed on private property.

One of the highlights is the Guggenheim mines, which turned a profit of $2,000 per day and established the Guggenheim fortune, later to become ASARCO. One of the Guggenheims' sons, Benjamin, perished in the sinking of the *Titanic*. The Little Johnny Mine was owned by the "Unsinkable" Molly Brown, who survived that calamity.

Headframe of the Robert Emmet Mine, about halfway through the self-guided tour.

Finn Town has two interesting stories. First, this is the headframe of the Robert Emmet Mine, a lead-zinc and silver producer. During the infamous miners' strike of 1896–1897, a fight broke out here between the strikers and the mine guards. Miners sought a raise after their pay was reduced from $3 per day to $2.50 due to the 1893 Silver Panic. The mine owners refused to pay a higher hourly rate, and during the impasse, nine out of every ten mines sat idle. As the miners grew more agitated, the owners brought in the state militia. After six months the miners relented and went back to work at their old wage, but many of the mines never reopened, as they had filled with water and mud. So in a way, both sides lost.

The second story involves the Guggenheims' Wolftone Mine, where the discovery of a rich lode of zinc in 1911 was cause for celebration. The owners brought in a fully catered meal, musicians, flowers, lighting, and waiters, then lowered 250 guests, six at a time, for a party at the 1,000-foot level that lasted several days.

At the Chrysolite Shaft, a man named "Chicken Bill" Lovell sprinkled stolen high-grade ore in the mine to get a good price when he sold out. The new owner, H. A. W. Tabor, sunk the shaft another 25 feet and encountered a zone worth $1.5 million.

At the Ibex Mines, superintendent J. J. Brown devised a method for draining the constantly flooding workings, and the owners were so happy, he was awarded a one-eighth share of the mine, rich in copper and gold. His wife was Molly Brown.

Most of these stops are doable with a passenger car or minivan, but the final stop takes you up a very steep drive to the Venir Shaft. The views are great, but the road isn't.

Some of the most well-known scoundrels, outlaws, and otherwise outsized personalities of the American West spent time in Leadville—"Doc" Holliday, "Soapy" Smith, "Big Nose Kate," Jesse James, and "Bat" Masterson, among others. You should make your way there, too.

49. Matchless Mine Tour

Land type: Historic mining district near Leadville
County: Lake
Elevation: 10,492 feet
GPS: 39.25431, -106.27053
Best season: Late summer
Land manager: Private
Material: Mine tour photographs; fine gold and small flakes at the panning station
Tools: Camera; pan provided
Vehicle: Any
Special attractions: Tabor House Museum
Accommodations: Full services in Leadville, including hotels, motels, B&Bs, hostels, and RV parks. Developed campgrounds nearby, such as those at Turquoise Lake.
Finding the site: From Seventh and Harrison in Leadville, drive east on Seventh Street for 1.2 miles to the sign on the left.

Prospecting

The Matchless Mine is a historic silver mine purchased in 1879 by H. A. W. (Horace) Tabor. Tabor had grubstaked August Rische and George Hook with the funds to keep exploring, and they used his front money to discover the Little Pittsburg Mine, kicking off the Colorado Silver Boom. Tabor sold his one-third interest (some say one-half) for a million dollars, then turned around and purchased the Matchless Mine for just over $100,000. The mine was fabulously wealthy, estimated to have produced $7.5 million during its peak operating years.

Tabor was elected the second lieutenant governor of Colorado, and was appointed to the U.S. Senate for a brief term.. When he married his mistress Elizabeth, aka "Baby Doe," in Washington, DC, the invitations were inscribed on solid silver. He and Baby Doe toured the United States, living it up, but they too fell victim to capricious ore structures and the silver market crash of 1893. When Tabor died in 1899, flags were lowered to half-mast out of respect for the "Bonanza King of Leadville."

There is still a lot of rehabilitation work to do at the Matchless Mine; their current project is to shore up the iconic headframe.

Tabor's dying wish was apparently for Baby Doe to hang onto the Matchless Mine, which she did, but it never paid off for her again. She became a hermit, hiding in a shack on the property, trying to make a profit from the base metals there. She was often seen around town in the freezing cold, with burlap bags on her feet, and was found frozen to death in her shack in 1935 at the age of 81.

The mine is now open for above ground tours, both guided and self-guided, and has a panning station where you can learn to pan and interesting exhibits that teach you about hard-rock mining. Bring cash—they do not accept credit cards or checks.

On the way to the mine, or coming back to town, stop in at the depot for the Leadville, Colorado & Southern Railroad at 326 E. Seventh St. Old Engine 641 sits on a pedestal, ready for a glamour shot. Check the website at www.leadville-train.com for their schedule; they offer two-and-a-half-hour

Railroads were an important part of the early Leadville mining boom.

round-trips to the Continental Divide. Once the local aspens have turned golden, it's a real treat.

The 12.5-mile-long Mineral Belt Trail was created as part of Leadville's environmental remediation effort. The trail uses abandoned railroad lines for hiking, biking, cross-country skiing, and general sightseeing through the district and offers a leisurely pace through the old mines. Most of the area is private property, so keep in mind that rockhounding and gold panning are off-limits. Bring your camera instead.

50. Point Bar Public Area

Land type: Riverbank
County: Fremont
Elevation: 6,786 feet
GPS: 38.47101, -105.87118
Best season: Late summer for low water
Land manager: BLM
Material: Fine gold, small flakes
Tools: Pan, sluice; dredging and high-banking require a permit.
Vehicle: 4WD recommended to avoid walking in
Special attractions: Arkansas River
Accommodations: Full services in Salida. Primitive camping on-site. Rincon Campground nearby on south side of river (38.47061, -105.86545).
Finding the site: There are two ways in, but when the water is high, you may be blocked from either way. The west option is to drive east from Salida on US 50. Starting from the intersection of Oak Street/CR 105 and Rainbow Boulevard/US 50, drive 4.8 miles east to CR 7. Turn left onto CR 7 and cross the bridge, then continue along the railroad tracks for a total of 0.9 mile. Turn right onto CR 45 and follow it for 2.3 miles. In 2017 this road was washed out, and you had to park about a mile before the coordinates where the road goes under the railroad tracks. To reach the site from the east, drive 11.3 miles from Salida, or another 6.5 miles. If coming from Cañon City, starting at the intersection of 15th Street and Royal Gorge Boulevard/ US 50, drive east 45.3 miles. Turn north onto CR 4/Howard Creek Road and cross the bridge, then follow CR 4 to its junction with CR 45, about 0.5 mile total. Don't end up on the Mustang Trail—it dead-ends. Take CR 45 west for 3 miles to Point Bar. Note that this road slides in between the river and the railroad tracks, and when the water is high (such as 2017), the river can cover the road. It gets extremely rough in places, and there is a tricky part at the railroad bridge over Badger Creek. You end up driving in Badger Creek about 100 feet toward the Arkansas River before picking up the road on the other bank where the brush parts.

Prospecting

This public area on the north bank of the Arkansas River provides excellent access to an area that is otherwise claimed or private. Most of the Arkansas

The road to and from Point Bar from the east runs right along the Arkansas River—a little exciting during the rainy season.

River is not available—there are either private claims, club claims, or private land to dodge. Sometimes the claim owners have replaced so many signs that unscrupulous prospectors tore down that they give up, so you don't really know the status until you call the BLM. They assured me that all the obvious spots on the Arkansas are claimed.

Point Bar is actually open all year, but you really need to work it when the water is low. The gold here is already very fine, so give yourself a chance to get more than the tiny flood gold that is common in this drainage.

There are a lot of rules and regulations to cover here. If all you want to do is pan or sluice, you don't need any special permits or authorization. Don't

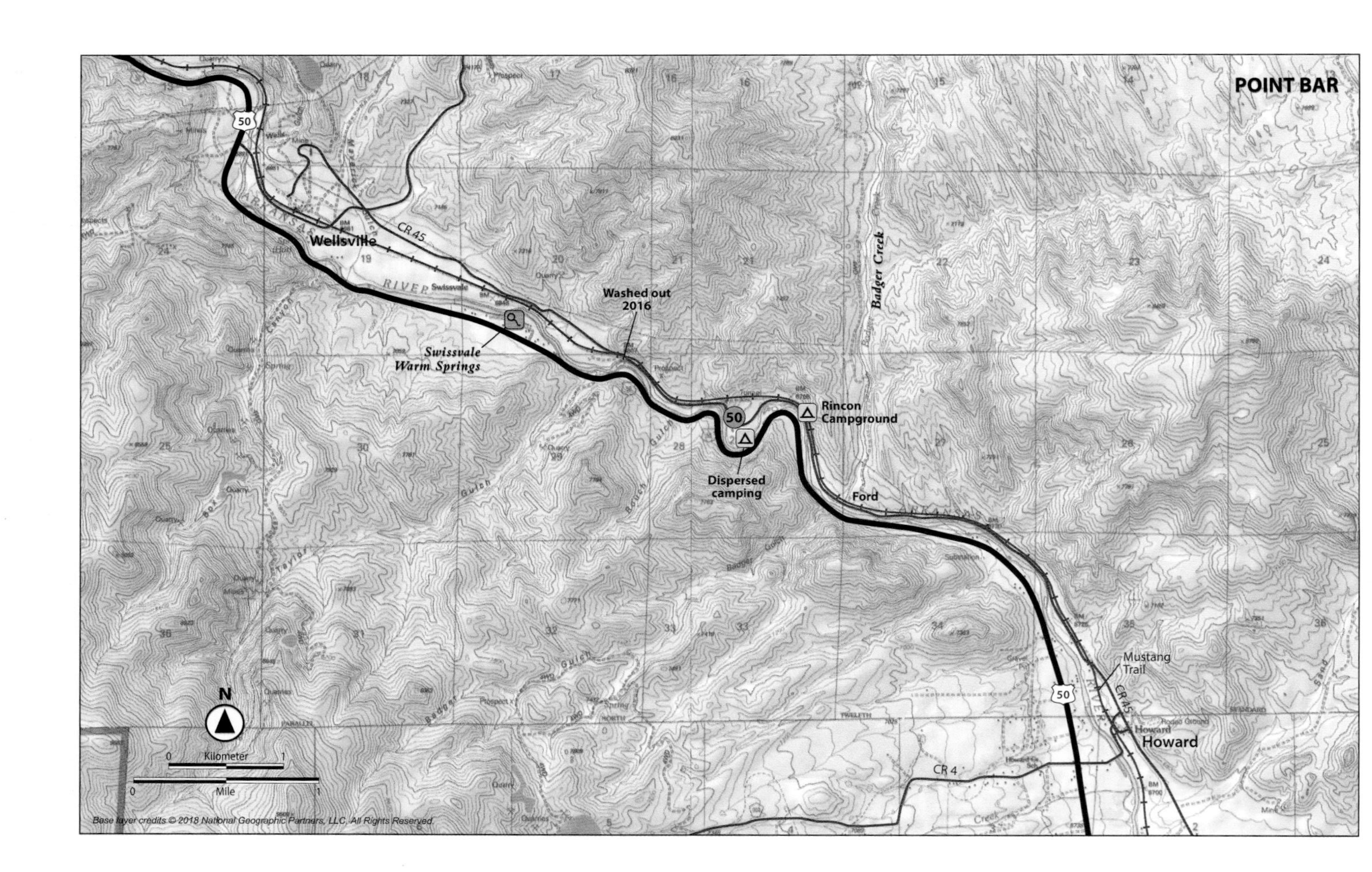
POINT BAR
Wellsville
CR 45
Swissvale
Warm Springs
Washed out
2016
50
Rincon
Campground
Dispersed
camping
Badger Creek
Ford
Mustang
Trail
Howard
CR 4
N
0
Kilometer
1
0
Mile
1
Base layer credits © 2018 National Geographic Partners, LLC. All Rights Reserved.

dig under trees, and don't dig "coyote holes" in the side of a hill, in the bank, etc. Those are pretty obvious by now—not causing a lot of damage is just common sense. Recreational panning includes battery-operated equipment such as 12-volt setups for your Gold Nugget Bucket, Power Sluice, Gold Cube, Gold Magic, or Camel system. If you want to high-banker or dredge, or perform any other activity that requires a gas-powered engine, you'll need a notice or permit from the BLM's Royal Gorge Field Office in Cañon City, or from the Arkansas Headwaters Recreation Area in Salida. The annual cost is $25 as of 2017. You can't anchor a dredge using cables that span the river, due to the high rafting traffic. Your dredge must be 4 inches or smaller at the nozzle. High-bankers should direct their runoff into buckets or big pans to let the water settle.

As far as primitive camping here, you are restricted to 14 days in a 45-day span. The restroom is available from April 1 through September 30; after that, you're expected to bring a portable toilet. Fires are permitted whenever fire season is open, but you must use a fire pan. There is no garbage service, so pack it in, pack it out. Don't drive off the roads.

You'll probably run into quite a few people in the busier summer months. Fortunately there is a little shade among the trees, but it gets very hot. It's a popular spot because while the gold is fine, it is plentiful. For once, you are located on the "correct" side of the river—the big bend is in your favor, so move up to where you can move bigger rocks and cobbles. Leave yourself plenty of time to explore, talk with fellow miners, and dig deep, and you should do well.

While I'm here, I should mention Badger Creek, which you forded when coming in from the east. There is a promising report about the creek at www.goldfeverprospecting.com/cogolo.html: "If you go 4 miles up Badger Creek, in the gravel bars and benches you can find placer gold, with copper minerals. This area was one of my favorites to prospect when I lived in Colorado." The road up the creek was gated at the railroad bridge, which was unfortunate. One likely spot is on FR 5965 at 38.51441, -105.85305. I just don't know if it's gated or private.

51. Lower Arkansas

Land type: Riverbank
County: Fremont
Elevation: 9,465 feet
GPS: 38.41073, -105.58397
Best season: Late summer
Land manager: BLM
Material: Fine gold
Tools: Pan
Vehicle: Any
Special attractions: Royal Gorge Bridge
Accommodations: Full services in Cañon City. Developed camping at Hayden Creek Campground and primitive camping on USFS land near Coaldale.
Finding the site: From Cotopaxi on US 50, drive 7.1 miles east to the small settlement of Texas Creek. If coming from Cañon City, drive about 26 miles west on US 50. The public spot is on the north side of the river, across the bridge over the Arkansas. Once over the bridge, turn right to the main parking area. The area to the north is a large off-highway vehicle (OHV) park. The fishing access spot on the left is private land, but it is commonly used by anglers and you'll often see cars parked there.

Prospecting

This spot west of Cañon City serves as the lowest easy gold-panning locale to try on the Arkansas River drainage. It is maintained as multiple use by the BLM, mostly for access to the Texas Creek OHV area. It's easy to find and fairly pleasant. There is a good little beach area just downstream from the bridge, and a hiking trail leads to another good spot.

Although lode mines line the cliffs on both sides of the Arkansas River valley, the gold by now is very fine, and you'll be hard-pressed to find much more than tiny colors. Also, don't be confused by the name Texas Creek. The community of Texas Creek is basically the cafe and RV park off US 50 at the bridge. The actual body of water known as Texas Creek flows into the Arkansas from the south and slows to a trickle in the summer. The topo maps show a couple of rock quarries on Texas Creek but no mines. The area to the

Fishing is common from this spot on the north bank of the Arkansas River at "the other" Texas Creek.

north of the river from the community of Texas Creek is drained by Texas Creek Gulch. There was a mine at 38.45473, -105.60823 and a couple more prospects below it. The Table Mountain jeep trail leads right through the mineralized area. Another big mine with many more tailings is at 38.47129, -105.58784.

While in the area, check out the Gold Mine Rock Shop at 44864 West US 50 in Cañon City (38.49599, -105.32441). They have a fantastic selection of rocks, minerals, prospecting supplies, rockhounding gear, maps, books, and more. It's a great spot to pick up something, ask a few questions, and recharge

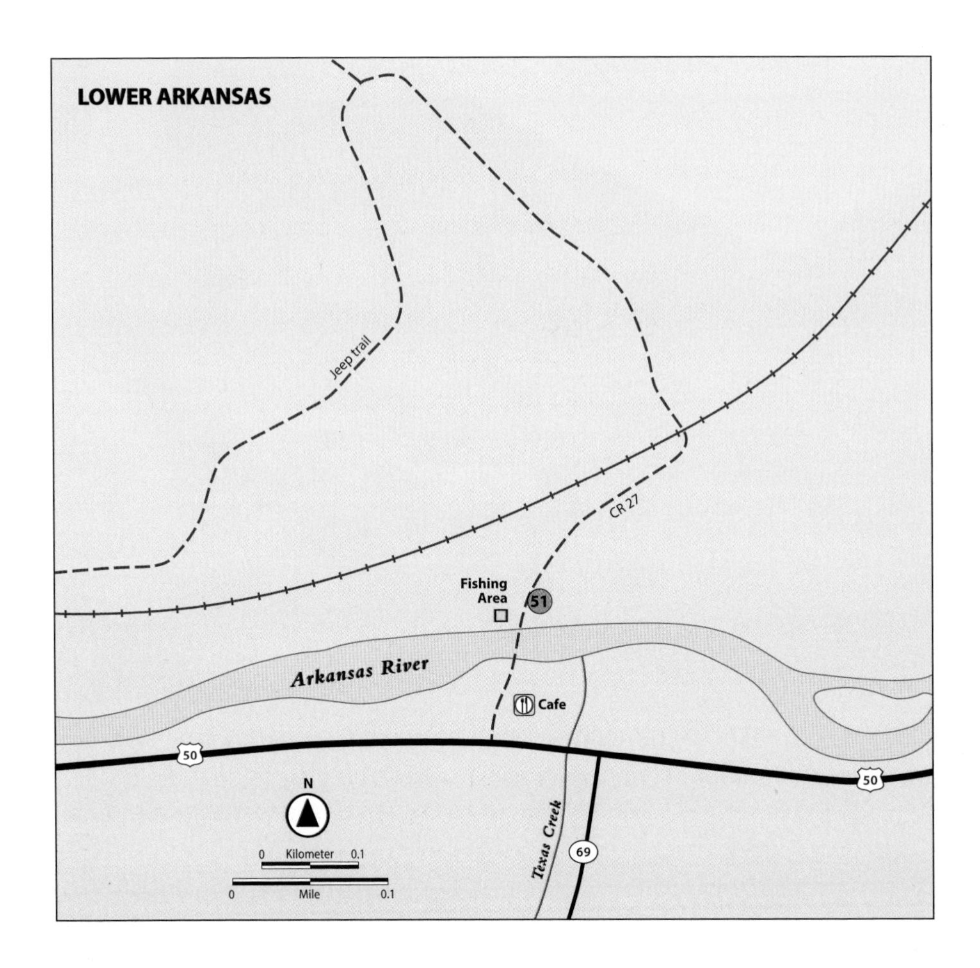

to head back into the field. One of the owners told me she sees people working in the river all the time, from Parkdale to Salida, but like everyone else, she recommended the BLM public area Cache Creek (Site 43).

52. Breckenridge

Land type: Mountain creek, riverbank
County: Summit
Elevation: 9,424 feet at Site A; 10,310 feet at Site D
GPS: A - Dredge: 39.53004, -105.99812
B - Tiger Gulch: 39.52125, -105.96346
C - South Fork Swan Road: 39.49483, -105.94744
D - Lower French Gulch: 39.48637, -105.98061
Best season: Late summer
Land manager: White River National Forest
Material: Fine gold, small flakes
Tools: Pan, sluice
Vehicle: Any; 4WD recommended for Sites C and D due to rough roads. I saw a minivan at Site D with a flat tire.
Special attractions: Downtown Breckenridge
Accommodations: Full services in Breckenridge, from resorts to B&Bs and motels. RV resorts include Tiger Run RV Resort. Public campgrounds are rare; try Dillon Reservoir north of town or Selkirk Campground on the other side of Boreas Pass. Dispersed, primitive camping near town is unlikely; toward Boreas Pass is your best bet.
Finding the site: From downtown Breckenridge at Lincoln and Main, drive north 0.6 mile to join CO 9 and head north another 3 miles. Turn east onto Tiger Road and drive 2.6 miles to turn left onto Dredge Loop and the parking area. The coordinates are from the parking area. To reach Site B, return to Tiger Road and drive up the drainage about 2.1 miles. Turn right onto Rock Island Road, and after 0.1 mile turn right onto Summit Gulch Road. The tailings at Site B are easy to spot to your left. To reach Site C, return to Tiger Road and turn right. Drive up along the tailings for 1.5 miles, then turn right onto FR 355.1B, which cuts through the tailings. Go about 0.9 mile to a collection of junctions. FR 355 heads southeast along the South Fork, and FR 567 heads southwest to American Gulch. A White River National Forest map will come in handy here to show private land. To reach Site D, the road does not go through, so you'll have to backtrack to town. From Lincoln and Main, drive north on Main Street to Wellington Road and turn right. Drive about 1.1 miles, then continue straight onto French Gulch Road. Drive 2.8 miles to the parking area where I took the coordinates.

Prospecting

In late spring and early summer of 1859, multiple mining parties swarmed the area that would become Summit County. According to author Rick Hague, the first recorded gold discovery was by Ruben Spaulding. Other sources credit the Weaver Brothers with multiple gold discoveries at Breckenridge; at Georgia Gulch, American Gulch, French Gulch, and Humbug Gulch on the Swan River; and on the Blue River at the confluence with French Gulch. In the early days the district was known as the Blue River Diggings, but in 1860 the post office was granted and named Breckenridge. It was named either for Thomas Breckenridge, one of the miners in the party of US General John C. Fremont, who explored the area in 1845, or for US Vice President John Cabell Breckinridge, who sided with the South in the Civil War. In the case of the latter, some accounts claim that the first *i* in the name was changed to an *e*, since Colorado was a Union state and Breckinridge chose the wrong side.

Either way, placer mining was the main activity during most of the early years. Harry Farncomb located the source of gold in French Gulch at Farncomb Hill in 1878, and by 1880 he opened the Wire Patch Mine. It was known for gold found as wires, leaves, and crystals, and is world famous for producing "Tom's Baby," a 156-ounce mass discovered by Tom Groves and Harry Lytton in 1887. Tom was said to have wrapped the precious bundle in a blanket and showed it off around town.

Hydraulic mines used monitors, elevators, and other contraptions as late as 1891, while lode deposits added to production, but output increased greatly in 1908 with the introduction of bucket-line dredges on the Blue and Swan Rivers (Koschmann and Bergendahl 1968, p. 116). Nine boats in all plowed through gravels in the valley. According to www.townofbrecken ridge.com, city officials were worried about keeping jobs going during the Great Depression, and permitted the Tiger #1 Gold Dredge to work from the north edge of town to the south end of Main Street. World War II halted gold mining nationwide, and the dredges fell silent. Of the 1 million ounces of gold produced in the Breckenridge District, about 750,000 ounces came from placer mining. In 1961 the first ski resort opened in Breckenridge; the town's character changed dramatically, and it retains little of the grit of an old mining camp.

Koschmann and Bergendahl (1968, p. 116) describe the wire gold at Farncomb Hill based on 1911 reports from Ransome (pp. 16-20) thusly: "The Farncomb Hill veins are remarkably persistent considering their narrowness.

The old Bucyrus dredge remains floating amid the tailings on the Swan River at Horseshoe Gulch.

They are rarely more than one-half inch wide, and some have been traced for distances of 300 feet. These veins are noted for their rich pockets of native gold which have supplied specimens of wire and leaf gold to museums and collectors throughout the world."

After checking the dredge ruins at Site A, work your way up Tiger Gulch. It would be nice if you could pan around the dredge, but you can't. Topo maps show the dredge tailings as continuous up Tiger Gulch, but it looks like there is a break at 39.52454, -105.98165 that is worth checking if it isn't posted. There have been good reports about Galena Gulch, but it rarely has water to pan. We collected some sphalerite and pyrite from the tailings at Site B after crossing the dredge tailings, but no good opportunities existed to pan there. This is mostly private land until you get to the junction of the Middle Fork and the South Fork near the old town of Parkdale. Keep going up past the Humbug dredgings and check above the tailings at Site C in all directions that aren't posted. Both forks are a bit swampy, so American Gulch to the southwest is your best bet.

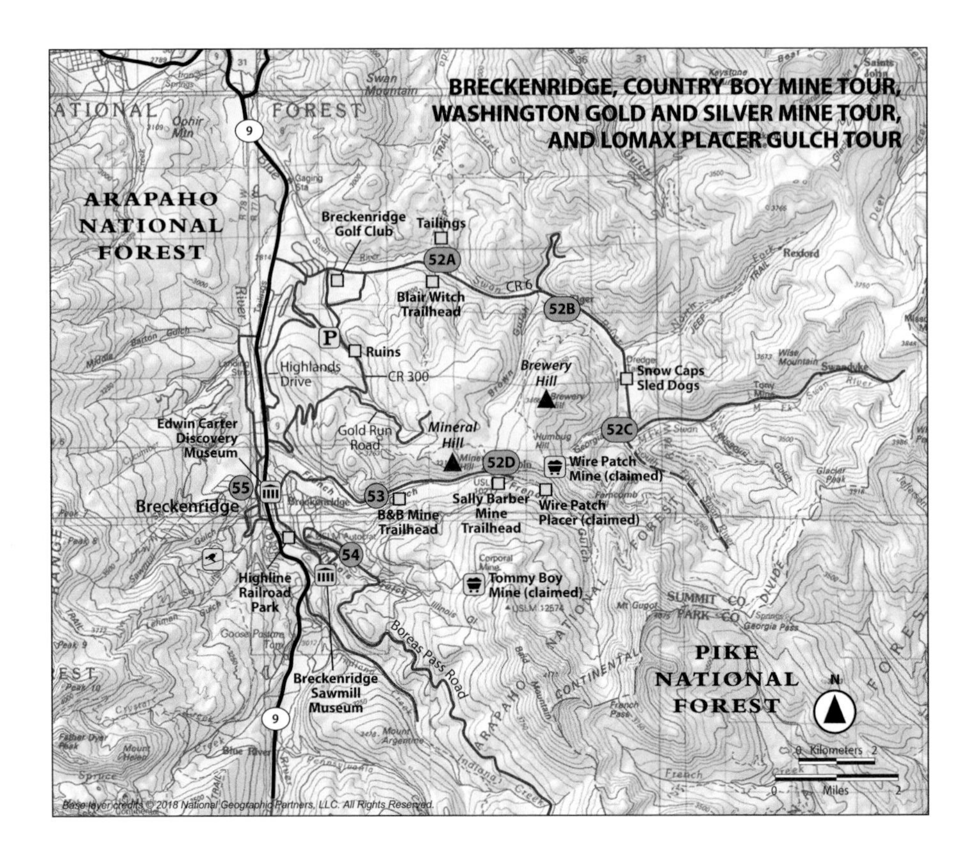

One interesting spot that I didn't get to personally investigate is up Gold Run. There is parking at 39.51491, -106.02647 and the ruins are at 39.51097, -106.02113, about a half-mile hike to the southeast. I can't tell if CR 300 is gated or if you can drive there.

The coordinates at Site D for French Gulch are easy enough to reach, but there is a gate, so you'll have to hike down to the creek to pan or hike in via the road to more of French Gulch. Private land is up there, and the Wire Patch Placer at 39.48371, -105.96851 and the Wire Patch Mine at 39.48748, -105.96601 are both claimed. Every local I asked about panning directed me to Site D, and you should be able to pan some tiny wire gold there with some patience.

For scenic views but no panning, try the drive up Baldy Road to the Tommy Mine headframe at 39.46398, -105.98758.

53. Country Boy Mine Tour

Land type: Mine tour
County: Summit
Elevation: 9,930 feet at mine entrance
GPS: 39.48091, -106.01438
Best season: May through Oct only. Closed during winter months.
Land manager: Private
Material: Fine gold, small flakes
Tools: Pan provided
Vehicle: Any
Special attractions: French Gulch
Accommodations: Full services in Breckenridge, from resorts to B&Bs and motels. RV resorts include Tiger Run RV Resort. Public campgrounds are rare; try Dillon Reservoir north of town or Selkirk Campground on the other side of Boreas Pass. Dispersed, primitive camping near town is unlikely; toward Boreas Pass is your best bet.
Finding the site: From the intersection of Lincoln Avenue and Main Street in Breckenridge, drive north less than 0.1 mile and turn right onto Wellington Road. Follow Wellington for 1.1 miles, then continue straight onto French Gulch Road. Drive 0.9 mile and you'll spot the sign for the turn to the right toward the Country Boy Mine parking lot at 542 French Gulch Rd.

Prospecting

From their website at www.countryboymine.com, I already knew the Country Boy was going to be a fun time. The tours have been running for over twenty years, lauded by the *New York Times*, the *Today Show*, the Travel Channel, and more. It was voted one of the top things to do in Breckenridge. As soon as I got out of the Jeep, I could hear happy children running around the parking area. This mine is totally family-friendly, with room for the kids to roam around, climb on structures that include a 55-foot ore chute/slide, and generally let off extra energy. The panning area is authentic, with pay dirt to scoop up and big metal pans like the old-timers used. There are ruins, mountains, mining, tailings—it's a great place.

The gift shop is well-stocked with books, gear, souvenirs, and more. The staff is helpful, friendly, and patient, and they know their business.

The panning area at the Country Boy Mine is an easy place to practice your technique.

The 45-minute tour itself is a treat, through well-lit but chilly underground workings—hard hat required! The tour guide will describe the work below ground, including a noisy live demonstration of a jackleg drill. As they explain, back when mining was just starting in the western United States, many of the workers hailed form the Cornwall tin mines in England. It seemed like every worker had a "Cousin Jack" who also needed work. Soon, they were all called Jack.

In the early days before power equipment took over, holes for blasting were driven by a man holding a steel chisel in one hand and swinging a hammer in the other. After each blow, the miner turned his chisel one-quarter turn. That was single-jacking. Double-jacking was when one miner held the chisel and another miner was then free to swing a bigger hammer. It took a lot of trust to hold your fist on that chisel as the sledgehammer came down, in conditions lit only by candles.

The powered jackleg drill was the next improvement, stationed with a steel boom to secure it to the top and bottom of the drift. If it ever came loose, it could kill or maim a man quickly. A slower death came from the dust created by the drill—it would lodge deep in a miner's lungs and cause silicosis. Soon, manufacturers figured out how to send a stream of water inside the drill shaft to apply to the dust, greatly reducing the hazards.

Once back outside, there is a petting station where kids can meet the family of miniature Sicilian donkeys, otherwise known as burros. Back in the early days before powered ore carts, burros were common in western mines for a brief time. They seldom left the mines, and their lot in life was difficult. However, they were quickly replaced by electric trams. Still, they make a charming addition to the tour.

The Country Boy Mine began in 1887. It produced mostly gold and silver to start with, and later yielded lead and zinc. There is a nice write-up of the local history on their website (lightly edited):

> *The Country Boy Mine is located in the famous French Gulch area, where gold was discovered by a miner named French Pete in 1860. This valley was abundantly rich in gold, silver, lead and zinc and played a meaningful role in the gold rush of Breckenridge. French Gulch is home to the* Reiling Gold Dredge *boat, built in 1908. It extracted gold-bearing ore and sank in French Gulch, where it can be seen [39.48378, -105.99557]. Tom's Baby was discovered in the same neighborhood, a 13.5-pound crystallized piece of wire gold! The owner, Tom Groves, wrapped the specimen in a baby blanket bundle and carried it around town like a baby. Also in the vicinity is the Wellington Mine, which generated millions of dollars in gold. It was the largest mine in the area and ran on and off until 1973.*

Tom's Baby is now on display at the Denver Museum of Nature & Science (Site 75).

54. Washington Gold and Silver Mine Tour

Land type: Mine tour
County: Summit
Elevation: 10,049 feet at parking lot
GPS: 39.46849, -106.02319
Best season: June through Sept only. Closed during the winter.
Land manager: Breckenridge Heritage Alliance
Material: Fine gold, small flakes
Tools: Pan provided and yours to keep
Vehicle: Any
Special attractions: High Line Railroad Park in Breckenridge (189 Boreas Pass Rd.)
Accommodations: Full services in Breckenridge, from resorts to B&Bs and motels. RV resorts include Tiger Run RV Resort. Public campgrounds are rare; try Dillon Reservoir north of town or Selkirk Campground on the other side of Boreas Pass. Dispersed, primitive camping near town is unlikely; toward Boreas Pass is your best bet.
Finding the site: From the intersection of Lincoln Avenue and Main Street in Breckenridge, drive south on Main Street for 0.4 mile, then turn left to stay on Main Street for another 0.3 mile. Turn left onto Boreas Pass Road and drive 1.2 miles. Turn right onto Illinois Gulch Road and drive a little more than 0.2 mile to the small parking area on the left at 465 Illinois Gulch Rd.

Prospecting

The first time I visited the Washington Mine, it was closed, but there are no gates and I was able to walk around and appreciate the old buildings and mine equipment at my leisure. This is a compact operation, run in conjunction with the Breckenridge Heritage Alliance, and there are plenty of exhibits, filled with old gear and memorabilia. So don't be tempted to just take your own tour and never go back. I returned and found that this is a delightful underground tour, with all the typical information about drilling holes and blasting dynamite. It's not as big and sophisticated as some, but the tour guides are very knowledgeable. Visitors will walk in for the underground portion and sit in the shade outside for a safety lecture and an informative talk on the local

Panning station and artifacts at the Washington Mine. Be sure to call ahead—this tour has limited slots and fills up fast.

Magnified wire gold specimen from French Gulch, which one of the tour guides at the Washington Mine keeps in his pocket.

history. This is just a small remnant of the larger operation that stretched up the hill, but it's a good taste of what lode mining was all about.

The panning station has good gold, stocked from nearby placer areas, and this is the only tour I visited in the state that lets you keep your pan. The tour guide will provide individualized instructions on panning, which to me was the strength of this tour.

You can order tickets online at www.breckheritage.com/mine-tours-gold-panning to schedule a tour at least an hour ahead of time. The site is open Tuesday through Sunday for tours at 11 a.m. and 1:30 p.m. It's best to plan ahead, as during the summer months the limited-capacity tours can fill up well in advance. Dropping in won't guarantee a spot on the tour.

55. Lomax Placer Gulch Tour

Land type: Mine tour
County: Summit
Elevation: 9,679 feet
GPS: 39.48329, -106.05271
Best season: June through Sept only. Closed during the winter.
Land manager: Breckenridge Heritage Alliance
Material: Fine gold, small flakes
Tools: Pan provided
Vehicle: Any
Special attractions: Edwin Carter Discovery Center in Breckenridge (111 N. Ridge St.)
Accommodations: Full services in Breckenridge, from resorts to B&Bs and motels. RV resorts include Tiger Run RV Resort. Public campgrounds are rare; try Dillon Reservoir north of town or Selkirk Campground on the other side of Boreas Pass. Dispersed, primitive camping near town is unlikely; toward Boreas Pass is your best bet.
Finding the site: From the intersection of Lincoln Avenue and Main Street in Breckenridge, drive west on Lincoln Avenue, which turns into Ski Hill Road. Drive about 0.4 mile up the hill, and look for the big sign on the left.

Prospecting

The Lomax Placer Gulch Tour is unlike the typical gold mine tours in Colorado, as it is not a lode mine—it's a placer operation. Instead of jackleg drills and ore carts, the Lomax Mine was the site of flumes, hydraulic monitors, ditches, and Long Tom sluices. It sits in a small gulch not far from downtown, now overgrown and built over with residences. But at one time, this was a major producer, employing an army of men who washed down the slopes and sluiced the runoff.

The placer material here is most likely related to the valley floor of the Blue River. Dredges operated from the north to as far south as Main Street; you can still see the tailings. Illinois Gulch and French Gulch drain into the valley from the west, and the Briar Rose Mine, at the top of Peak 10, drains into Sawmill Gulch, which may have been shifted a bit north in the past. However the gold reached the Lomax, it was plentiful. The tour guide can

Entrance to the Lomax Placer Gulch Tour.

The panning area at Lomax Placer Gulch is right on the same creek that miners used to work.

show you an old picture from the early 1900s of the Lomax area completely devoid of trees. It has rejuvenated itself since then.

Since there are no underground tours, the highlight at the Lomax is the panning area in the gulch below the various cabins and buildings. After you complete your self-guided tour of the structures and appreciate the mining memorabilia in the various exhibits, the tour guide will get you started panning in the gulch. They don't have to "salt" their panning area, as plenty of color constantly washes down from the gulch to this day. That gives this tour a much more casual feel—since there are no scheduled trips underground, you can drop in. Open Tuesday through Sunday from 11 a.m. to 4 p.m. Plan at least one hour for panning. A full guided tour with a knowledgeable local expert takes place on Saturdays at 3 p.m. See www.colorado.com/factories mine-toursgold-panning/lomax-placer-mine for more information.

56. Montezuma

Land type: Riverbank
County: Summit
Elevation: 9,498 feet at Site A; 10,572 feet at Site C
GPS: A - Power lines: 39.60522, -105.91581
B - St. Johns Mine ruins: 39.57054, -105.88132
C - Snake River: 39.56404, -105.86069
Best season: Late summer
Land manager: White River National Forest
Material: Fine gold, small flakes
Tools: Pan, sluice
Vehicle: 4WD required for Site B or for any additional exploring.
Special attractions: Montezuma is still an interesting town, with limited services. It has suffered through five fires, so many of the original buildings are gone.
Accommodations: Nearest full services are in Dillon, and there is camping there, too. Dispersed camping along upper Deer Creek and Peru Creek is possible; park well off the road.
Finding the site: From Silverthorne on I-70, take exit 205 for US 6/CO 9 toward Dillon. After exiting, follow US 6 for 7.8 miles, then take the Montezuma Road exit. Follow Montezuma Road for 1.8 miles to the easy parking spot on the south side of the road. You can probably find additional access to the river along here—just be careful to dodge private land. To reach Site B, drive on to Montezuma, about 3.7 miles, then turn right onto CR 275/St. Johns Road. Follow this road for 1.4 miles until you reach the ruins. To reach Site C, backtrack to Montezuma and resume traveling south on Main Street/Montezuma Road for 1.3 miles to the large turnaround area.

Prospecting

These locales barely scratch the surface for Montezuma. Before you reach town, you can turn on CR 260 and follow Peru Creek up as far as you like; there is access to the creek, and the Maid of Orleans structure is about 1.4 miles up. The Pennsylvania Mill complex starts at 39.60039, -105.81333, another 2.5 miles farther. A lot of remediation work has been done up here, fixing the mine waste seepage with holding ponds. You could also go all the

Site A along the upper Snake River, looking downstream toward Montezuma. Tailings from the Superior Mine are on the left.

way to the eastern flanks of Ruby Mountain on CR 260 and see a lot more mining ruins.

Another good side trip would be to take Third Street/CR 264 out of Montezuma to the north and zigzag up Santa Fe Mountain to the old Quail Mine at 39.58039, -105.84581. There is some private property close to town to dodge on the way up, but plenty of tailings and old ruins can be found along this rough road.

Site A offers easy access to the Snake River, after it is joined by Peru Creek. This site is commonly used by anglers for access and had no claim markers. Some of the other easy access spots, such as near the bridge, are on private land, as is the tailings area below Grizzly Gulch. Peru Creek is a little more open.

Site B at Saints John is more of a photo opportunity right now, as there has been reclamation on the ruins and signs are posted warning you to stay off the vegetation. Topo maps show the Wild Irishman Mine above the St.

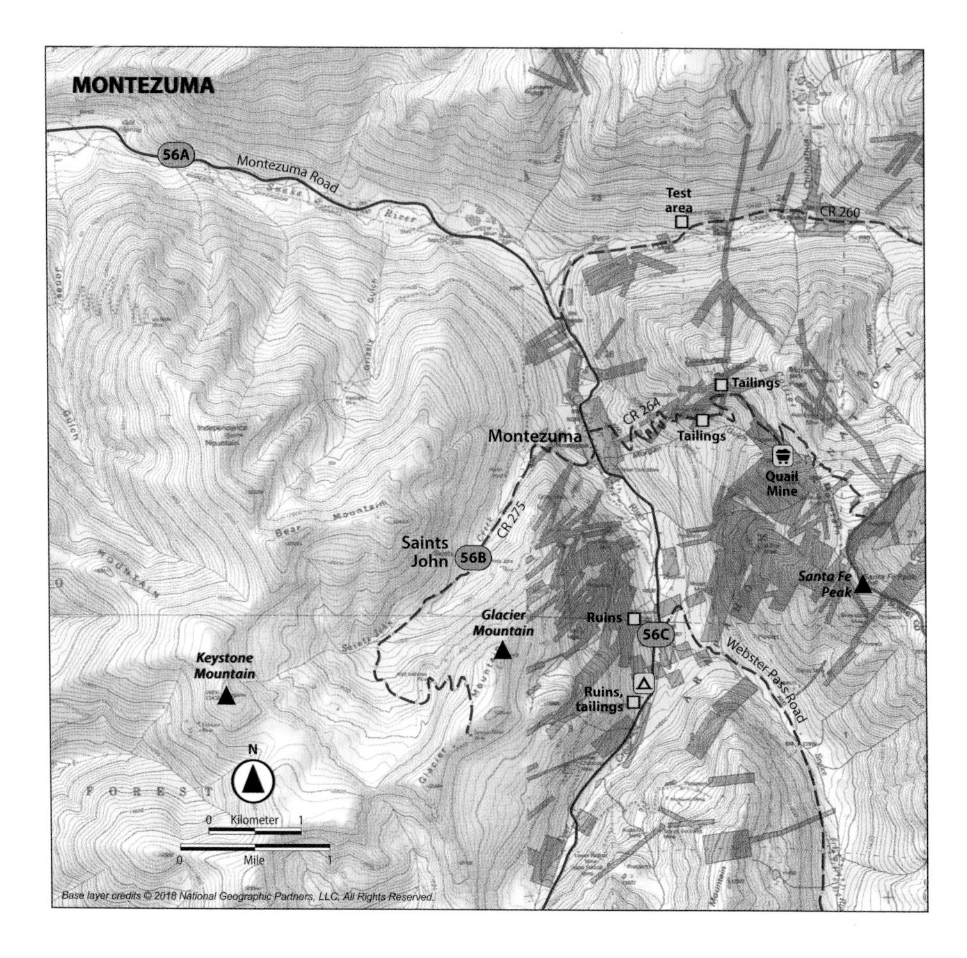

Johns ruins, and it appears to be open for rockhounding. There is also private property up here, so obey the signs. St. Johns was originally named Coleyville, for John Coley, who discovered silver here in 1863. It was said to be one of the earliest silver strikes in the territory. Freemasons had a strong influence in this area, and in 1867 they renamed the town Saints John, for their patron saints John the Baptist and John the Evangelist (Dallas 1985, pp. 174–75). There was a post office here from 1876 through 1881, and the town boasted a strong moral fiber, thanks to the Boston investors who backed the mine. According to Dallas, the town had a 350-volume library, but no saloon. The high elevation brought severe winters and tough cooking conditions. Dallas said the town had a saying: "Beans for Sunday dinner had to be started by Friday afternoon." Old topo maps show the road from Saints John going all the way through to Breckenridge, by the way.

At Site C there is a big turnaround/parking area where Deer Creek enters the Snake. Below the culverts are big cobbles and good colors. This is *not* a camping spot. If you drive farther up the mountain, the first right takes you to the Superior Mine, but be on the lookout for activity there. Several mines were located farther up Deer Creek; the other side of the Continental Divide is the headwaters of the North Fork of the South Platte River. Prospectors coming over the mountains from Georgetown founded Montezuma in 1865.

Be cautious about speeding through Montezuma, as they are serious about enforcing their speed limits.

57. Alma

Land type: Riverbank
County: Park
Elevation: 10,995 feet at Site A
GPS: A - Magnolia Mill: 39.35737, -106.08316
B - Buckskin Joe Mine: 39.29012, -106.09654
C - Old arrastra: 39.29432, -106.10469
D - Paris Mill: 39.29482, -106.10589
E - Buckskin Creek: 39.30802, -106.11802
F - Mosquito Creek: 39.27939, -106.12946
G - London Mill: 39.29496, -106.15204
Best season: Late summer
Land manager: Arapaho National Forest and Pike National Forest
Material: Fine gold, small flakes; tailings, ruins
Tools: Pan
Vehicle: 4WD suggested
Special attractions: The tiny Alma Mining Museum; Bristlecone Pines
Accommodations: Full services in Breckenridge and Fairplay; limited services in Alma. Primitive campground at Kite Lake, at the top of Buckskin Creek.
Finding the site: From Breckenridge, Alma is about 15 miles south on CO 9. It is about 5.5 miles north of Fairplay. From "downtown" Alma at Buckskin Street and Main Street, drive north on CO 9 for 4.8 miles, then turn left onto CR 4. Drive 0.9 mile, then make a slight right on PV32 and continue for 0.9 mile. The Site A ruins are around Montgomery Reservoir. To reach Site B, return to Buckskin Street in Alma and turn west on CR 8. Drive about 1.8 miles, past the Boston & New York Placer sign, and look for a left turn. The ruins of the Buckskin Joe Mine are about 0.4 mile away. To reach Site C, resume traveling up the mountain on CR 4. About 0.9 mile up, past the geodesic domes, there is a little waterfall and the outline of an arrastra in the creek bed. To reach Site D, keep going another 0.1 mile. There is a wide turnout here. At some point in the future, this will be the entrance to the Paris Mill, listed here as Site D, but it is still undergoing renovations right now. About 1.1 miles farther, there is access to Buckskin Creek at a ford (Site E). Just 0.3 mile farther up is the Sweet Home Mine, which used to offer tours but is no longer open to the public. To reach Site F, backtrack to Alma. Drive south on CO 9 for just

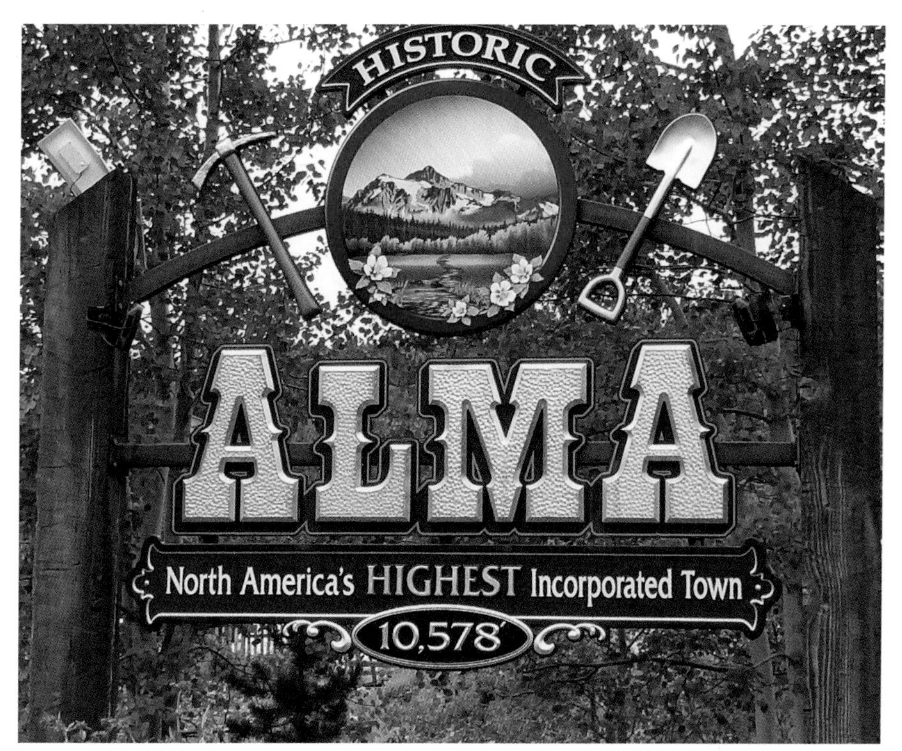

Alma is a charming little town of 235 residents with a rich mining history. There are fun opportunities to pan and explore in several directions, to the north, west, and south.

0.2 mile, then turn right onto CR 10/Park Hill Street. After 1.5 miles make a slight right onto Mosquito Pass Road and drive 2.7 miles. Turn left onto Burke Road, and after 0.1 mile locate the creek passing under CR 696 (which you could have also taken). To reach Site G, return to Mosquito Pass Road and drive 1.7 miles farther up the mountain.

Prospecting

Alma saw lots of mining in almost every direction from town. The biggest producer was the Orphan Boy Mine, located at 39.27788, -106.10185 and detailed in H. Court Young's book, *The Orphan Boy: A Love Affair With Mining*. Court helped restart the mine after World War II, but the property is again shuttered. In all, lode mines in the district produced 1.2 million ounces of gold (Koschmann and Bergendahl 1968, p. 109), with another 28,000 ounces from placers, mostly on the South Fork of the Platte River, which is completely locked up in private land and active claims.

With the Platte off-limits, turn your attention to Mosquito Creek and Buckskin Creek. At Ghosttowns.com, there is a little history behind Buckskin Creek's namesake and the long-faded camp that also bears his name:

> *This camp was named for Joe Higgenbottom who was named "Buckskin Joe" for the leather clothing he wore. Legend has it the mine was found while a man named Harris was deer hunting. He shot at his prey and thought he hit it but the deer ran off. In searching for the bullet or traces of blood, he found the bullet and it had lodged itself in an outcropping of placer gold. More than 5,000 people showed up in this town by 1861. By 1864 the town was deserted and the mother lode had played out. Ranches continued in the area for a long time and some of the buildings at this town were carted off to nearby Fairplay.*

Site A is the picturesque Magnolia Mill, and a quick pan in the creek near it should be OK. There is more mineralization up here, such as near Quartzville at 39.34485, -106.07763, so feel free to explore.

Site B is on famed Buckskin Creek, but there is a lot of private land, and nothing is left of the actual town except the cemetery. The mine has been open in the past to look over, but that could change so beware.

Site C is an old arrastra foundation. Large stones set in the circle ground up ore, and the creek washed the lighter material away. It's not common to see one of these. This one may have had a burro pulling the top weight. I listed it for the arrastra, but it's on the creek and you could try a pan or two.

The Paris Mill at Site D is currently owned by the Chiwawa Mining Co. It was built in 1895 to support the Paris Mine on Mount Bross, started twelve years earlier. A tram connected to the Paris Mine and Hungry Horse Mine, but the mill ran in fits and starts, and after 1935 it was more of a showpiece, with its ball mill, stamp mill, and rod mill all intact. In 2004 it was listed as endangered, and while the creek runs right through here, it's probably all private land.

Site E is private above the ford, but there were no claim posts downstream in 2017. It sits just below the famed Sweet Home Mine, source of some of the finest deep red rhodochrosite crystals in the world, including the Alma Rose, the hit of the 1966 Las Vegas Gem and Mineral Show and the star attraction at the Rice Northwest Museum of Rocks and Minerals in Hillsboro, Oregon. An even larger specimen, the Alma King, is on display at the Denver Museum of Nature & Science. Closed in 2004, the Sweet Home Mine is no longer open to the public.

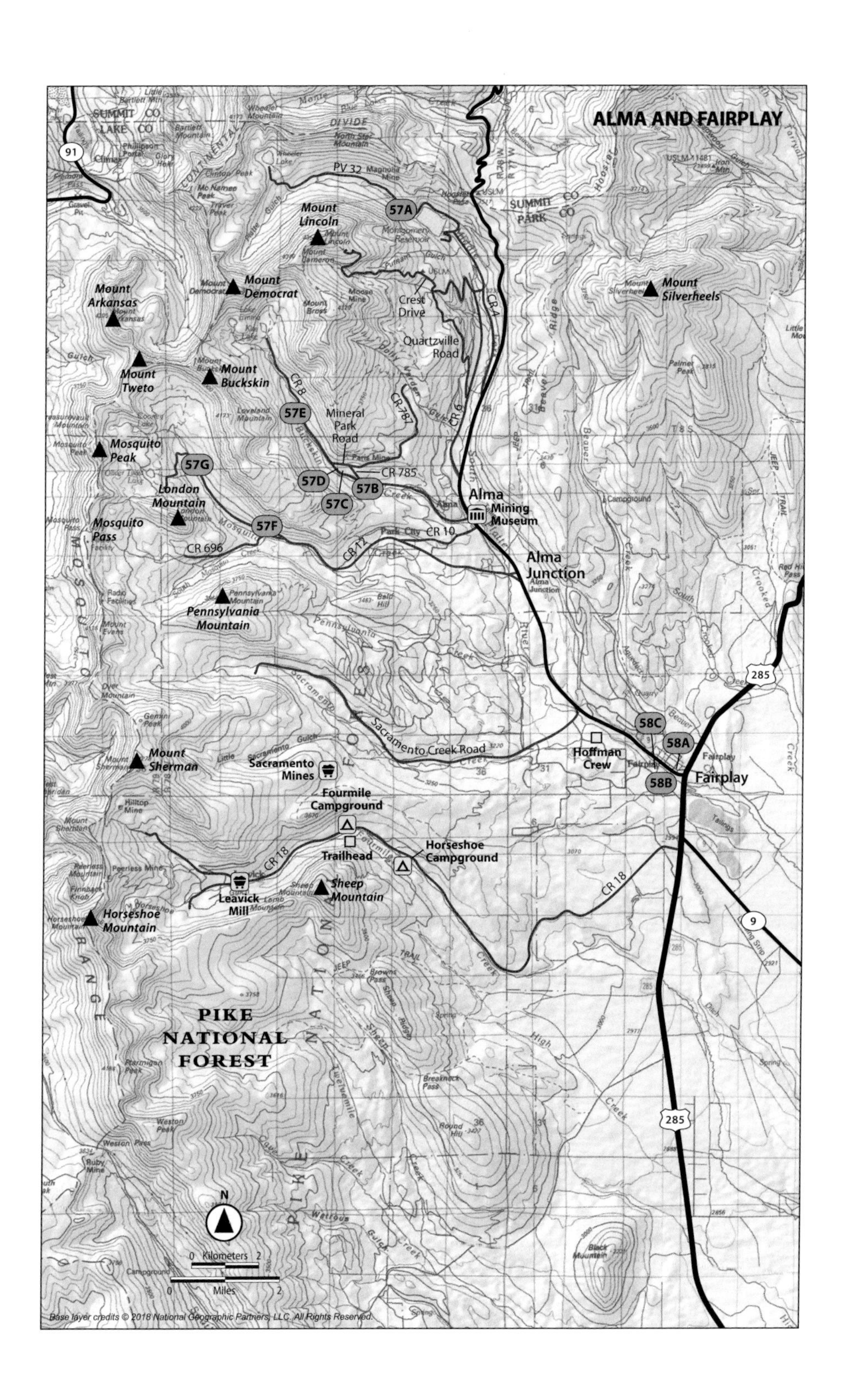

Base layer credits © 2018 National Geographic Partners, LLC. All Rights Reserved.

Over on Mosquito Creek, Site F is OK for panning, and Site G is the London Mill; you could pan downstream from the mill.

These many coordinates only scratch the surface around Alma. If you want more 4WD adventure to visit old mines, try up CR 448, which leads to the Paris Mine at Mount Bross. The coordinates are 39.29978, -106.10319. There are several other sites or ruins along Mosquito Pass Road. The American Mill is at 39.28280, -106.14182. The North London Mine supplied the London Mill by tram, and it is located at 39.28996, -106.16237. The ruins of what I believe was the Hock Hocking Mine are at 39.26699, -106.09981, off CR 698, which leads to Pennsylvania Mountain.

Note that this area now has a self-guided tour starting in Alma, with nine historic markers explained in a handy brochure. Check at City Hall at 59 E. Buckskin Rd. in Alma. For more information about the Paris Mill, visit the South Park National Heritage Area website at www.southparkheritage.org/paris-mill.

Finally, a quick note about Pennsylvania Mountain, which yielded the biggest gold nugget in Colorado. It's all claimed, which you'd know if you watched a few episodes of *Gold Rush*, when famed Colorado mining-legend Freddy Dodge took Todd Hoffman up there to scout around. Parker (2009, p. 43) has an extensive write-up that will leave you yearning to give it a try. Two of the largest alluvial nuggets known in Colorado (see Site 75) came from Pennsylvania Mountain: the "Turtle Nugget," a 7.8-ounce nugget shaped like a turtle found in 1990, and a nugget found in 1937 that weighs 12 troy ounces, or 1 pound. The top of the mountain is gravel, which is odd for an area 12,000 feet up. I hoped to find some open ground on Pennsylvania Creek or Mosquito Creek where the mountain gravels might have eroded to the north, but I didn't have any luck with that bright idea.

58. Fairplay Beach Public Area

Land type: Riverbank
County: Park
Elevation: 9,910 feet
GPS: A - Visitor center: 39.22151, -105.99671
B - Public area first entrance: 39.22021, -106.00112
C - Second entrance: 39.22102, -106.00373
Best season: Apr through Oct
Land manager: City of Fairplay
Material: Fine gold, flakes
Tools: Pan, sluice
Vehicle: Any
Special attractions: Reconstructed mining camp and museum (39.22516, -106.00365, Fourth and Main). Several old mines at the top of Four Mile Creek Road.
Accommodations: Full services in Fairplay. Camping on Four Mile Creek Road at Horsetooth Campground.
Finding the site: From the intersection of CO 9 and US 285, drive north on CO 9 about 0.1 mile to the visitor center to get your permit. Parking is on the left, on the north bank of the Platte River. You'll have a view of the dredge tailings there. To reach the public panning area, drive back to US 285 and turn left (west) on US 285. Drive 0.4 mile across the river, then take the first major left onto Platte Drive. Go about 0.5 mile to the first entrance to the panning area. It's now gated, but there is parking here. The other parking spot is less than 0.2 mile farther up the river on the right.

Prospecting

In the earliest days of the western gold rushes, a new camp had two common themes: greed and generosity. If the first miners on the scene were greedy and took all the best land for themselves, a camp would often be called "Hog 'Em" or "Grab All." But if the first division of claims spread the wealth around, the name "Fair Play" might stick. In the case of this old mining camp, a group of miners, not at all happy with their findings along the eastern fringe of the Rockies, decided to look elsewhere. They found a stream not far from the South Platte River that offered good possibilities. Deciding to stay, they named the town Tarryall. News leaked out, and others came to try their luck.

They were met with firm resistance and told to keep on going "or else." They did so and located the deposits on the South Platte, which they divided equitably, according to the write-up by Harry Chenoweth at Ghosttowns.com.

Prospectors flooded into the South Park area after gold was discovered there in 1859. The South Platte is lined with dredge tailings, and whatever isn't private is claimed. The Fairplay Beach area is open to the public with a modest day pass or seasonal pass, available at the Town Hall, the Hand Hotel on Front Street, High Alpine Sports at Sixth and Main (where you can also pick up some prospecting equipment), or in advance via the town website at http://fairplayco.us/docsforms/2017_Gold_Panning_permit_application.pdf. The site originated with the help of multiple grants and cooperation between the town of Fairplay, Park County, Colorado Parks and Wildlife, the Denver Water Board, corporate sponsor US West, and the Upper South Platte Water Conservancy District. They set aside the dredging spoil zone and turned it into a fishing hole with restrooms, paved trails, and picnic tables, all connected to the town of Fairplay by a footbridge and a trail network. Once constructed, the site was turned over to Fairplay's city government, which manages the area.

There are many rules to abide by. Dredging and high-banking are not allowed; you also can't bring your favorite battery-powered contraption. Pans, sluices, and shovels are the rule. No digging upstream from the footbridge, and no digging into the banks of the river, no matter how tempting those big cobbles look. Yes, there are some preexisting holes there—avoid them. You'll do fine just moving big rocks. If you can get a good hole going on the right side of a river bend, you should do well. Note that like most public areas, the spots closest to the parking area are hammered the hardest. If you visit more than once, you might consider exploring across the footbridge and downstream. The public area actually goes all the way to the highway, and maybe one in a hundred panners make it that far. If you can visit when the water is lowest, you might do OK out in the tailings, especially if you spot an abandoned channel formed when the river was running high. But your best bet is to stick to the edges of the riverbank. Note that you can also park on the north side of the public area at 39.22175, -105.99891 and walk down.

Some of the "flood gold" you get at the surface could have been dislodged by the Hoffman operation that you see on the Discovery Network's *Gold Rush* show. Todd Hoffman left the Yukon in 2016 to try a district in eastern Oregon that was closer to home, but that didn't work out, so he teamed up with Freddy Dodge to operate on some property between Alma and Fairplay on the South Platte. Their operation isn't open to the public, but I've known

Public panning beach at Fairplay, from Site B, looking back toward the gate at Site A. Note the footbridge, which offers access to the other side of the river. The open area starts below the footbridge.

Todd since the first season when I interviewed him for *Gold Prospectors* magazine, and he told me they got into some pretty good gold eventually. Todd gets a lot of scrutiny and second-guessing on his mining operations, but the show has been a consistent success.

If you have time, check out Fourmile Creek—there is another historic auto tour up that drainage to Horseshoe Basin. The Leavick Mill is at 39.19513, -106.13727. The Sacramento Mines are at 39.22217, -106.10895, via Thompson Park Road.

Also near Fairplay the famed Snowstorm dredge sits in a gravel pit above the river. It is the last intact floating dragline gold dredge in Colorado, and it could be the last in the nation according to an article by Kenneth Jessen, a noted Colorado historian. Built by the Bodinson Manufacturing Company in San Francisco and used intermittently from 1941 to 1976, it had 10,000 cubic yards of capacity, used a 150-foot boom, could reach 70 feet in depth, and weighed about 700,000 pounds. The dredge worked the river gravels between Fairplay and Alma, processing 600 tons of pay dirt per day, and produced an estimated 12,400 ounces of gold. There is talk of restoring it and creating a tourist attraction, but right now it is on private land and not open to the public. Go to www.savethesnowstorm.com for more information about Gold Unlimited and the many partners attempting to purchase and relocate this historic treasure.

59. Tarryall Creek

Land type: Riverbank
County: Fremont
Elevation: 9,806 feet at Site A
GPS: A - Como Roundhouse: 39.31583, -105.89176
B - Selkirk Campground turn: 39.37054, -105.95107
C - Michigan Creek: 39.41085, -105.88384
Best season: Late summer
Land manager: Pike and San Isabel National Forests
Material: Fine gold, small flakes
Tools: Pan, sluice
Vehicle: 4WD recommended; Boreas Pass Road is very rough in places.
Special attractions: Como Roundhouse; Boreas Pass
Accommodations: Full services in Fairplay. Camping at Selkirk Campground and Michigan Creek Campground.
Finding the site: From the intersection of US 285 and CO 9 in Fairplay, drive east on US 285 for 9.3 miles. Turn left onto CR 33/Boreas Pass Road and drive 0.5 mile. The roundhouse information kiosk is on the road to the right. To reach Site B, drive farther up Boreas Pass Road for 6.7 miles, past the old Roberts Cabin and up the steep road. Turn left onto FR 801/CR33A and go back down a couple of switchbacks for 1.3 miles to turn right onto CR 801/FR 406. Go about 0.1 mile to the turn for the Selkirk Campground. To reach Site C on Michigan Creek, return to US 285 and turn left, then drive 6.7 miles. Turn left onto Michigan Creek Road and drive 2.9 miles. Make a slight right to stay on Michigan Creek Road and drive another 2.5 miles. Turn left to stay on Michigan Creek Road; the turn to the campground is in 0.6 mile. The maps show Michigan Creek Road going all the way over Georgia Pass to the South Fork of the Swan River out of Breckenridge.

Prospecting

Named by a group of miners who found good workings and decided to stay for a spell, the Tarryall Placers produced about 67,000 ounces of gold, mostly from 1859 to1959 (Koschmann and Bergendahl 1968, p. 110). There are few lode mines in the area. Dredging spoils dominate one stretch along Boreas Pass Road above Como, and more tailings are located along US 285 east of Como. These are all posted as private land.

Sometimes these giant culverts are really tempting sampling targets. This one is on Michigan Creek.

The small town of Como was a rough and violent place in its beginning. Chinese gangs battled Italian immigrants in the city streets, and a mine explosion killed more than twenty men in 1893. One old tale recalls how the Reynolds gang robbed a stage in 1864 and stole the gold box, only to be chased down by an angry posse. Six men were killed in the gunfight that ensued during the gang's capture, and six more were apparently shot before trial when they wouldn't reveal the whereabouts of the gold, which was never found (Dallas 1985, p. 50).

For railroad buffs, Como was an important part of the state's railroad history. Trains went over Boreas Pass to Breckenridge in 1882, and by the late 1880s ten to twenty-six trains passed through town daily, according to www.southparkheritage.org. High altitude, long winters, strong winds, and

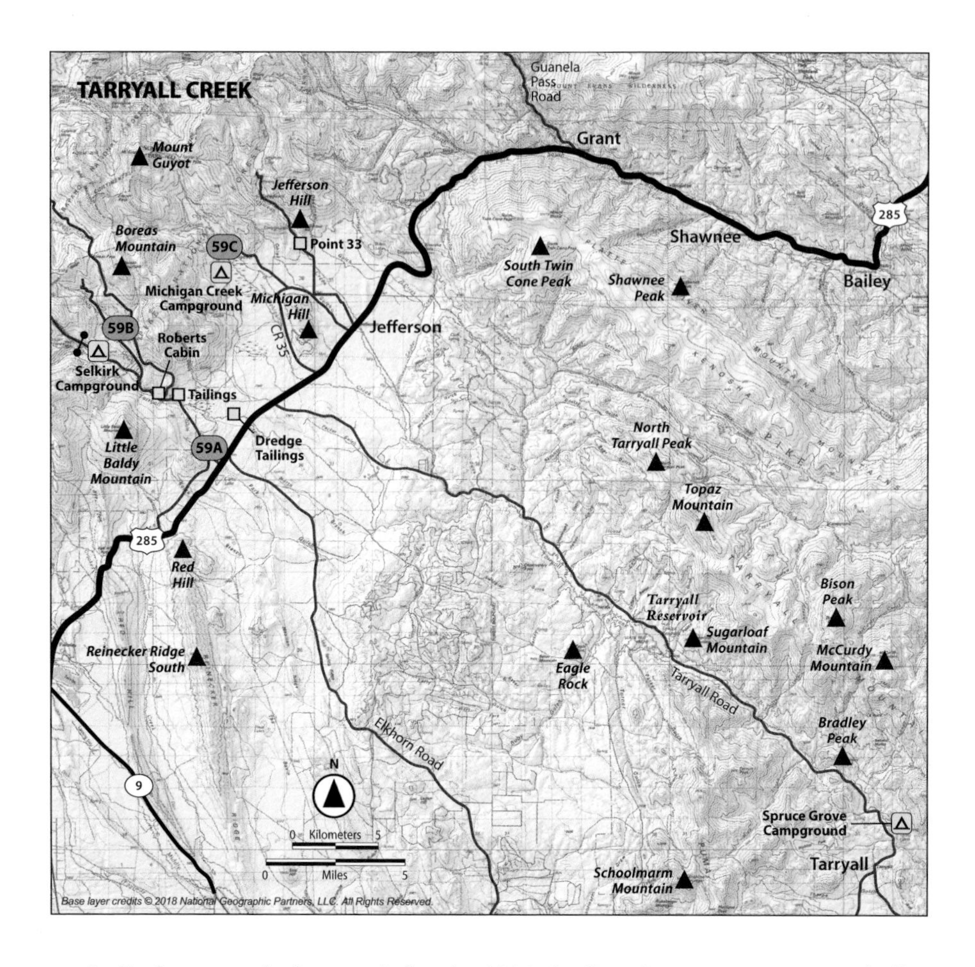

periodic fires were bad enough, but in 1910 the line from Gunnison to Chalk Creek Canyon closed, as the Alpine Tunnel was too challenging to keep open. By 1937 the route to Fairplay was shuttered, too. The roundhouse was built by Italian stonemasons and is open for tours Friday through Sunday, 10 a.m. to 4 p.m., during the summer.

Boreas Pass Road stretches to Breckenridge and passes the old town site of Dyersville, built around the Warriors Mark Mine. It was founded in 1881 and named for Father Dyer, a Methodist preacher who was noted by local miners for his ability to use a pair of forked sticks to feel the vibrations of the universe to "dowse" for ore deposits. Dyer quickly soured on the rough-and-tumble life of the miners in his town and moved to Fairplay (Dallas 1985, p. 70), where he met with even rougher handling. Local toughs shaved his horse and tried to steal his church. "If Hell is any worse than Fairplay, I certainly want to be saved from going there," he muttered (p. 76).

Most of the best areas for placering are locked up as private land and gated. Purgatory Gulch, Montgomery Gulch, French Gulch, Deadwood Gulch, and Australia Gulch, all feeding into upper Tarryall Creek, are gated just above the old Fortune Placer Mine, west of the old train station of Peabodys. All of Parker's advice (2009, pp. 43–44) is mouthwatering but not useful with the new gates that restrict access. Selkirk Gulch at Site B is usually dry, so look for access to North Tarryall Creek starting at the entrance to the campground and working upstream. Nearby Michigan Creek at Site C isn't nearly as rich as Tarryall Creek, but there are small colors.

Some decent spots can also be found lower down the Tarryall drainage. Starting at the parking area at 39.30561, -105.75068, look for access where you don't have to cross any fences or jump a claim, and continue south about 20 miles, past Tarryall Reservoir to Spruce Grove Campground, just north of what's left of Tarryall on CR 77. You might find some good crevices at the campground to clean out.

ipple Creek

Land type: Creek banks, historic town
County: Teller
Elevation: 8,837 feet
GPS: 38.71261, -105.18241
Best season: Late spring through fall
Land manager: BLM
Material: Fine gold
Tools: Pan, sluice
Vehicle: 4WD recommended if road is wet or icy. "The Shelf" is particularly nasty, even when hot and dry; yield to uphill traffic and keep an eye out for turnouts in one-lane sections.
Special attractions: Cripple Creek District Museum (38.74685, -105.17221); casinos in downtown Cripple Creek; visitor center and train depot (38.74742, -105.17249); Florissant Fossil Beds National Monument; Cleveland and Marsh dinosaur quarries
Accommodations: Full services in Cripple Creek. Primitive camping in the middle of the Shelf Road area between Cripple Creek and Cañon City. Wye Campground east at Penrose-Rosemont Reservoir and Mueller State Park on CO 67 north of Midland.
Finding the site: From Hartsel on US 24, drive east on CO 9 for 30.5 miles and turn right at Florissant onto CR 1. Follow CR 1 for 17 miles. You are now on Carr Avenue; go another 0.6 mile and turn right onto Second Avenue, which soon turns into CO 67 as you drive south. After 0.3 mile turn right onto Xenia Street/CR 88, also known as Shelf Road. Public land starts about 2.3 miles total from CO 67 and continues for about 4 miles to the "Keyhole." This road goes all the way through to Cañon City.

Prospecting

The Cripple Creek District was once Colorado's premier gold producer, and it was the second largest in the United States, behind the Homestake Mine in the Black Hills. It has yielded about 22 million ounces of gold so far, and is still going strong. The historic district itself is mostly taken over by thriving casinos, but when you visit the town, be sure to go east on Bennett Avenue to Carbonate Street and check out the railroad depot and visitor center.

"Keyhole" in the limestone formations along Shelf Road. Once you reach public lands, there are plenty of opportunities to sample the creek.

Here's the discovery story from Dallas (1985, 58):

For ten years, Robert Womack, a cowboy and a prospector, has been telling saloon bums in the dives of Colorado City that there was a fortune to be made in gold at Cripple Creek, the yard-wide stream that ran through his cow pasture. For a time, when news of a gold find sent scores of prospectors to the area, it appeared Womack might be right. But the claim turned out to be a salted mine; thus whenever mining men were tempted to listen to Womack, they recalled the Mount Pisgah Hoax, as it was known, and put their money into the newly discovered silver mines at Creede.

Womack eventually located a few small outcrops but soon sold his claims. Winfield Scott Stratton grew interested and determined that what the prospectors thought was the gray lead ore galena was actually a rare gold ore, a telluride named sylvanite (Koschmann and Bergendahl 1968, p. 118). Stratton dug for fifteen years and finally found gold on the Fourth of July, so he named

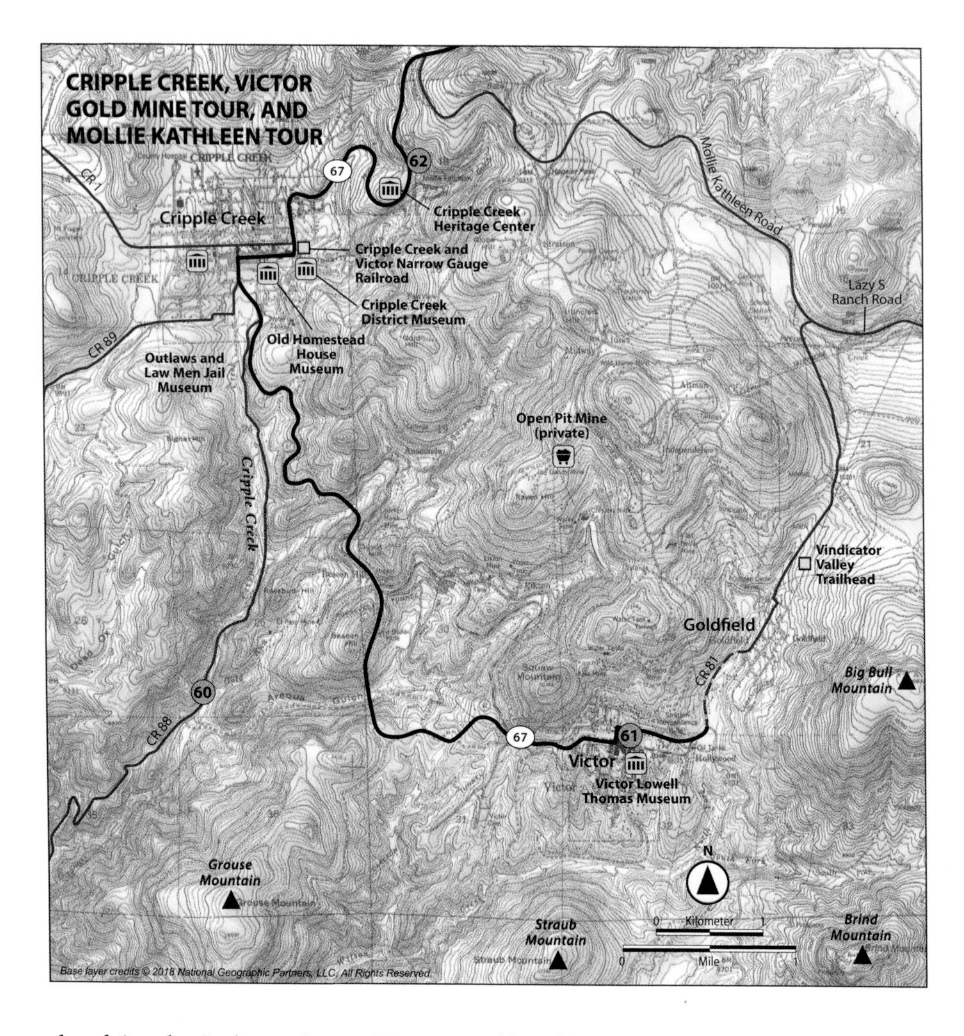

the claim the Independence. He eventually sold it for $11 million. From 1890 to 1910 more than 22 million ounces of gold were extracted from 500 mines in the Cripple Creek District. During the richest years, the mines created thirty millionaires.

The geology of the area is complex, and it has been studied by famous geologists such as Waldemar Lindgren (1906) and A. H. Koschmann (1949). They describe a compact area of 7 square miles, forming a volcanic caldera, with Miocene volcanic breccia about 30 million years old, sitting on Precambrian granite, gneiss, and schist at least a billion years old. Volcanic pipes and dikes cut the rocks. Fissures are filled with veins of quartz and fluorspar, with coarse pyrite, multiple tellurides, sphalerite, galena, and tetrahedrite making up the ore minerals. Some of these minerals are on display at the Cripple Creek District Museum (www.cripplecreekmuseum.com).

Cripple Creek itself doesn't show many scars or big fields of tailings. There are a few mines at the top of the road after it leaves the highway, with some interesting cribbing and some stone ruins at the Rosebud Mill, Brodie Mill, and Morning Glory ore house. I already knew about Shelf Road before I started any of the mine tours at Cripple Creek, but I asked all around and to a person, everyone recommended panning out here below the private holdings. It's mostly open BLM land, with well-marked private areas. Don't cross any fences!

You can easily access the creek from the coordinates south for 4 miles; eventually you start getting into posted land again, so beware. You'll probably see tents and campers during the summer months. The water flows pretty well through here, but even at this site, where the creek is just a trickle, you'll get color just below the ruts that ford the creek to a gate and fence. Once you get to the switchbacks as the creek slithers through the canyon, there are plenty of parking spots to choose from, so be sure to park safely.

61. Victor Gold Mine Tour

Land type: City streets, mine tour
County: Teller
Elevation: 9,713 feet at museum, where tour starts
GPS: 38.71004, -105.14011 at museum
Best season: Memorial Day through Labor Day only; closed during the winter.
Land manager: Private; Newmont Mining Corporation
Material: Fine gold in panning trough
Tools: Camera; pan provided
Vehicle: Any
Special attractions: Victor Lowell Thomas Museum; Victor's Gold Camp Ag & Mining Museum; vast hiking and biking trails through the district
Accommodations: Full services in Cripple Creek. Primitive camping in the middle of the Shelf Road area between Cripple Creek and Cañon City. Wye Campground east at Penrose-Rosemont Reservoir and Mueller State Park on CO 67 north of Midland.
Finding the site: From Cripple Creek, at the intersection of Carr Avenue and Second Street, take Second Street south; it turns into CO 67. Drive for 4.6 miles, around the mine, and into Victor on Victor Avenue to the cross street of Third Avenue. The Victor Lowell Thomas Museum is the starting point for the tour.

Prospecting

Newmont Mining Corporation (NYSE: NEM) completed their acquisition of the Cripple Creek and Victor gold mine in August 2015 for $820 million plus a 2.5 percent net smelter return royalty on potential future gold production from underground ore. Just writing about it takes me back to my days as an overeager reporter for the *North American Gold Mining Industry News*, a short-lived 1980s newspaper that covered the wild junior mining stock scene out of the old Vancouver Stock Exchange. The catchphrase in Vancouver was "It's not for widows and orphans!" For $3,000 a year, a mining company could sign up to publish all the news they might supply us, and most of them could only dream about advancing from their meager drilling programs to a giant operation such as this.

An expansion project is under way at Victor to combine cyanide heap leaching of the surface ore trucked out of their giant pit with a separate mill

Looking down into the great open pit of the Victor Mine. Note the drilling rigs on the lower left for scale. The mine is now starting to exploit richer ore from underground.

to process high-grade ore from underground. The leaching operation began in 1995, and the new mill is already in production.

Because the nearby mines used low-grade ore to pave the streets of Victor, the town was often referred to as "The City with Streets of Gold." Thanks to numerous mines visible from downtown, Victor was also known as "The City of Mines." It was said that a smart miner could walk 5 miles underground from Victor to Cripple Creek by using the vast underground tunnel network. Many adits and tunnels are now visible in the open pit.

The tour begins at the Victor Lowell Thomas Museum in Victor. Thomas (1892–1981) was an internationally acclaimed radio personality, writer, and world traveler, best known for publicizing T. E. Lawrence as "Lawrence of

Arabia" in 1918–1919. The museum has an excellent gift shop and a decent panning station out on the sidewalk. All proceeds go to the museum; as of 2017, the tour price was an extremely reasonable $8.50, conducted with an air-conditioned short bus and a knowledgeable local guiding the way. You'll enjoy excursions for photographing, viewing the crusher, climbing around on a huge ore truck, and more. Tours can fill up quickly, and they request that you call ahead at (719) 689-5509 during summer months. Check their website at www.victorcolorado.com/mining.htm for late-breaking news.

The city of Victor has teamed up to create a vast network of hiking and biking trails that "Tour the World's Greatest Gold Camp." Here's a link to the map: www.victorcolorado.com/stcfg/mapsign2017web.pdf. For example, the Independence Mill is at 38.71338, -105.13432. Additional photographic opportunities abound, but the lands here are mostly private, so no gold panning or rockhounding is allowed (yet—we can always hope). A lot of rehabilitation work is still planned for many of the historic headframes that dot the hills around Victor, and this is another good example of Newmont trying to do right by the community. They have invested heavily in environmental protection to minimize permanent scars to the land and are doing everything they can to be model corporate citizens at Victor.

62. Mollie Kathleen Gold Mine Tour

Land type: Old mine
County: Teller
Elevation: 10,002 feet at parking lot
GPS: 38.75355, -105.16052
Best season: Late spring through fall only. Closed during the winter.
Land manager: Private
Material: Photographs
Tools: Camera
Vehicle: Any
Special attractions: Cripple Creek Heritage Center, "across the street" from the mine tour
Accommodations: Full services in Cripple Creek. Primitive camping in the middle of the Shelf Road area between Cripple Creek and Cañon City. Wye Campground east at Penrose-Rosemont Reservoir and Mueller State Park on CO 67 north of Midland.
Finding the site: From the intersection of Second Street and Bennett Avenue in Cripple Creek, drive east on Bennett to the edge of town, where the main road becomes Carbonate Street, about 0.3 mile. Continue on Carbonate Street as it becomes CO 67 and proceed 1.5 miles north, then turn right onto CR 82/Mollie Kathleen Road. The entrance to the mine is on the right. The street address is 9388 CO 67, Cripple Creek.

Prospecting

Unlike most mine tours in Colorado, the Mollie Kathleen tour doesn't go straight into the mountain—it goes straight down, about 1,000 feet, in a small steel cage. During busy stretches, the tour runs every 20 minutes, but even then, you may be crowded in next to perfect strangers for the 2-minute descent. The ride down is on what's called a "man-skip" and is actually slower than an elevator, so don't worry about surviving a bone-jarring plummet through darkness. You'll don a hard hat (mandatory) and a heavy coat (a very good idea—temperatures plummet to 50°F /10°C). Part of the tour is on foot, and part of it involves riding on the electric tram. You'll see several minerals in

Like most tours, part of the Mollie Kathleen Mine underground tour includes a demonstration of the jackleg drill. This is the only tour in Colorado that takes you 1,000 feet straight down to get started.

the mine walls, including gold ore. No pets, backpacks, baby strollers, walkers, crutches, or wheelchairs are allowed; use a front-loading baby pack. There are no restrooms down below. Steel-toed boots are not required, but the ground can be muddy, so open-toed shoes are discouraged. For more information, check their website at www.goldminetours.com.

Here is the discovery story from their website, lightly edited, about how a bold woman came to locate her own mine:

> *In the spring of 1891 Mollie's son Perry arrived in the Cripple Creek area employed as a surveyor assigned to map mining claims of this country's newest and overlooked frontier. With all news of the day focused on Cripple Creek's gold, Mollie loaded the family wagon with supplies and joined the next wagon train heading west up Ute Pass to visit her son . . . In September of that year (1891), Perry was surveying upper Poverty Gulch when he saw a huge herd of elk. Later he told Mollie of the herd so she headed out to see for herself. As she made her way up Poverty Gulch (three hundred yards past Cripple Creek's first gold strike—Bob Womack's Gold King Mine) she grew winded*

and decided to rest. Looking downward as she caught her breath, Mollie noticed an interesting rock formation that winked back at her. Using a rock to break off a sample, she could hardly believe her eyes. The outcropping was pure gold laced in quartz. With her heart racing, Mollie nonchalantly hid the gold samples amongst her clothing. She had to be calm, as there were several prospectors in the area. She realized Bob Womack had overlooked her find for more than a dozen years, prospecting an area he had nicknamed Poverty Gulch.

Mining continued at the Mollie Kathleen until 1961, when the Carlton Mill closed and there was no easy way to process their ore. Rather than close the mine, the operators invested in upgrading the facilities to turn it into a public tour. You'll learn a lot about mining in the old days, with a good overall history of the district. In the back of your mind, you might be thinking you're strong enough and tough enough to work down there, but don't be fooled. Famed heavyweight fighter Jack Dempsey, born in Manassa, Colorado, in 1895, once signed on to work at the Portland Mine and then at the Black Diamond Mine in Cripple Creek. Back in those days, the mucker had one of the hardest jobs in the mine, and worked a 12-hour shift for $3 per day or less. Dempsey had a rough-and-tumble childhood, quitting elementary school to work and leaving home at age 16. He started visiting saloons and bellowing, "I can't sing and I can't dance, but I can lick any SOB in the house!" He rarely lost those barroom brawls, which usually only ended when one man could no longer stand. Dempsey only lasted one day as a mucker at both mines, getting fired for not being able to keep pace. He quickly returned to boxing.

HONORABLE MENTIONS

The following Central Colorado locales did not make it into the book, but they're worth a visit if you're looking for new areas to explore.

Cebolla Creek

Mostly rockhounding at 38.31944, -107.09689. White Iron had good tailings at 38.33462, -107.20256.

Gerrard

The Silver Dollar Mine at 37.79866, -106.51736 and the Monon Mine at 37.80142, -106.53893 both looked interesting on the maps.

Grape Creek

Located near Cañon City, at 38.40733, -105.32604. Dry, and some private land to dodge.

Grayback Gulch

Located at 37.57947, -105.28679, near the Denver Placers and Spanish Gulch at La Veta Pass, it's all private land.

Placer Creek

West of Salida. No idea what's here, but I had three coordinates to try: a lower spot at 38.57573, -106.16718; a middle at 38.58419, -106.16788; and an upper spot at 38.58854, -106.17864.

Sawmill Gulch

Above the Arkansas near Granite. Two Bit Gulch at 39.11182, -106.29462 and Sawmill Gulch at 39.11536, -106.29869 both looked worth a peek.

Turret

There might be some open roads around here.

Westcliffe/Silver Cliff

Too dry and too much private land for a gold-panning book, but might be worth checking. The Defender Mine at 38.15361, -105.45222 looks like it has good tailings.

Part III: Northeast Colorado

63. Georgetown Loop Tour

Land type: Mountain creeks
County: Clear Creek
Elevation: 8,743 feet at Georgetown; 9,136 feet at Silver Plume
GPS: A – Georgetown Devil's Gate Depot: 39.70163, -105.70651
B – Silver Plume Depot: 39.69447, -105.72462
Best season: Late Apr through Sept only. Closed in the winter.
Land manager: Private
Material: Fine gold
Tools: Pan provided
Vehicle: Any
Special attractions: Downtown Georgetown has excellent shopping; Georgetown Rock Shop (Sixth and Taos); Georgetown Heritage Center. Empire is a neat little town not far away.
Accommodations: Full services in Idaho Springs, Golden, and Denver. Camping options on Guanella Pass Road south of Georgetown. Dispersed camping on USFS land.
Finding the site: From Golden, drive west on I-70 for about 32 miles to exit 228 for Georgetown. Cross back over the interstate and turn left onto 15th Street. Go through the traffic circle to Argentine Street, then go 0.5 mile and stay right to continue on Loop Drive for about 0.8 mile. You'll reach the depot and the parking area for Site A. To start at Site B, take exit 226 for Silver Plume. The ramp ends at Woodward Street; go south under the interstate to Mountain Street, turn left, and stay left onto Railroad Avenue. The parking area is just ahead. If coming from the east on I-70, use the same exits to start your tour in either Georgetown or Silver Plume.

Prospecting

The Belmont Lode in the Argentine District was discovered in 1864 at the heads of Leavenworth Creek and Stevens Creek. The district is underlain by schist and gneiss of the Idaho Springs Formation and intruded in the Precambrian by Silver Plume granite. Later, in the Tertiary, porphyry dikes cut through the area, bringing quartz, fluorite, and carbonate veins carrying galena, pyrite, sphalerite, chalcopyrite, silver, and gold. About 25,000 ounces of gold came from this area, with a lot more silver. The Georgetown-Silver Plume District has about the same geology as the Argentine District, but

Start of the Georgetown Loop Railroad and Mining Tour, which combines railroads and mining for an unbeatable combination.

produced about 145,000 ounces of gold, with zinc another prominent commodity. There are two vein systems here—one carries gold with pyrite, while the other is mostly lead and zinc (galena and sphalerite), with lots of associated silver but little gold. Silver ores include hessite and argentite, and one mine contained platinum and iridium (Koschmann and Bergendahl 1968, p. 96; Lovering and Goddard 1950, pp. 141–42).

Georgetown is commonly known as the "Silver Queen of Colorado." It was founded in 1859 by a pair of Kentucky prospectors, brothers George and David Griffith; the older brother got top honors and the town is named for him. When James Huff discovered silver at Argentine Pass in 1864, the rush was truly on. Soon there were thousands of mines, prospects, and claims in the surrounding mountains. Georgetown was incorporated on January 10, 1868, when Colorado was still a territory, and it soon grew larger than nearby Idaho Springs, prying the county seat away. Georgetown secured the courthouse and prospered again when the Colorado Central Railroad laid tracks up the canyon from Golden in the 1870s. The only narrow-gauge tracks remaining from that expansion are the Georgetown Loop, which the tour uses between Georgetown and Silver Plume. The Colorado silver boom of the 1880s pushed the population of Georgetown to 10,000 citizens and brought Georgetown to equal status with other leading silver towns, such as Leadville. Locals actively agitated briefly to move the state capitol from Denver.

The town has currently shrunk to about 1,000 people, and the Georgetown Loop Historic Mining and Railroad Park is the main attraction. You can take

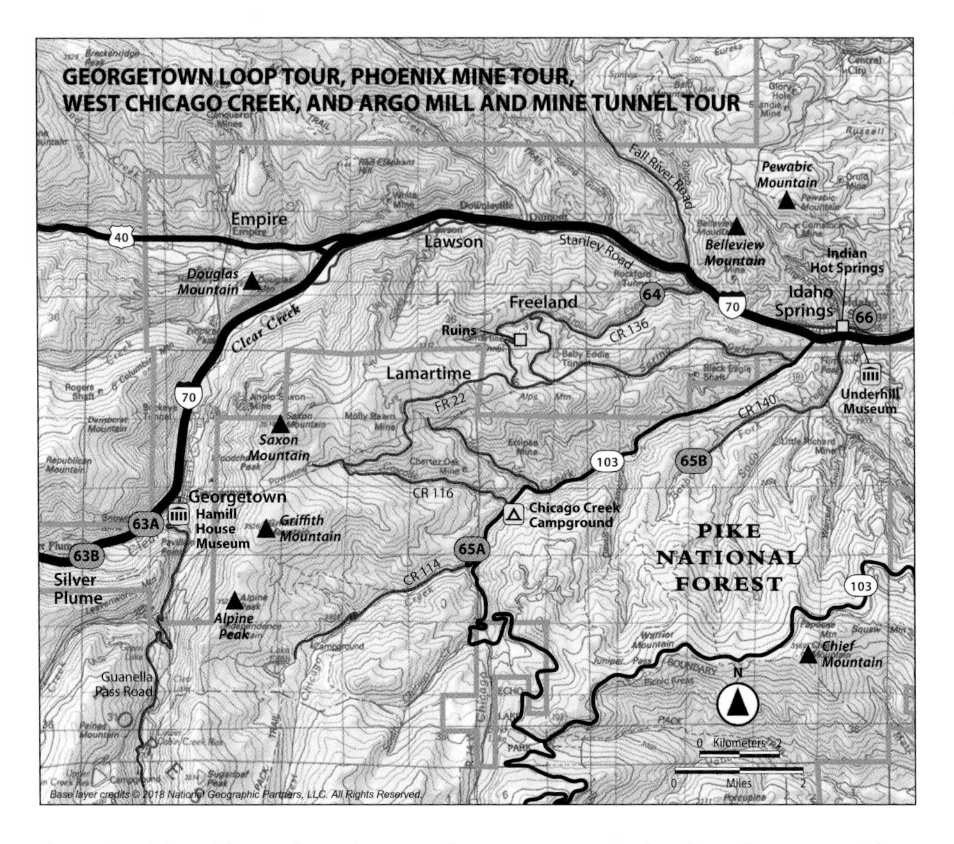

the train ride without the mine tour, but you cannot take the mine tour without the train ride, so prepare to spend about two and half hours, split evenly between the two activities. You can start the tour at either end of the run, so I included both coordinates.

Mine tour options include the Lebanon Silver Mine tour, with discussions of early-day mining, and the Lebanon Extension Mine tour, with more mineralogical features such as stalactites, calcite dams, and silver pearls. Your best bet is the Everett Gold Panning and Silver Mine tour, which covers basic early mining techniques and puts you at the panning station. It also includes a light lunch, and is the best bet for children. Check the website at www.georgetownlooprr.com/mine-tours-and-gold-panning for more information about weather policies, walking distances, attire, etc.

There are three classes of railroad cars to choose from: coach class, in an open-air boxcar with bench seating; parlor cars, which are fully enclosed and carpeted, with tables and chairs; and the ultimate, the Presidential/Waldorf Car, with a dedicated attendant and a "golden ticket" to take home.

64. Phoenix Mine Tour

Land type: Mine
County: Clear Creek
Elevation: 8,165 feet
GPS: 39.74862, -105.56581
Best season: Late spring through fall primarily, but winter tours are possible if weather permits.
Land manager: Arapaho National Forest; private mine
Material: Fine gold
Tools: Pan, sluice
Vehicle: 4WD required
Special attractions: Underhill Museum in Idaho Springs showcases the history of the Idaho Springs area.
Accommodations: Full services in Idaho Springs and Golden. Camping options are limited; try Guanella Pass Road/CR 381 to the west, such as the campground at Clear Lake.
Finding the site: From the west end of Idaho Springs, drive west on Colorado Boulevard to the end of town, but don't merge onto I-70. Instead, turn left onto Stanley Road. (You'd have to exit I-70 using exit 239 if coming from Golden to the east; if coming from the west, you'd still need to take exit 239 and backtrack.) Once on Stanley Road, drive about 1.1 miles and turn left onto Trail Creek Road. Follow it for 0.8 mile to the mine on the right, and park safely above the tour staging area.

Prospecting

In 1861 miners discovered the first veins of lode gold on Trail Creek, and after 1868, when the smelter at Black Hawk began operating, production from the area increased quickly. When the railroad arrived in 1879, another boost kicked in, but from 1910 to 1933 production generally declined. When the price of gold increased in 1934, there was another spurt of activity, but that too was short-lived. Koschmann and Bergendahl (1968, p. 95) include the Phoenix Mine as part of the Freeland-Lamartine District. They report about 200,000 ounces of gold produced in this small, 4-square-mile area, mostly from the nearby Freeland and Lamartine Mines. Both mines appear on old topo maps with abandoned towns bearing their names.

Crowding around the panning instructor at the Phoenix Gold Mine.

The bedrock geology here is dominated by the Idaho Springs Formation, a collection of gneiss and schist rocks dating to the Precambrian. Quartz diorite, granite, and pegmatites cut through the metamorphic rocks. More porphyritic intrusives punctured the country rock in Tertiary times and brought in the ores. Two types of mineralization are of interest to miners: pyrite-gold concentrations and sphalerite-galena concentrations. The gold is associated with the pyrite, and there is native gold here as well.

At the Phoenix Mine, owner-operators Al and Dave Mosch do a great job of educating the public, walking visitors through the usual tour standbys such as mucking, drilling, and processing. This is still a working gold mine, with a new section of the mine exploiting the Resurrection Vein, which is yielding over 2 ounces of gold per ton. They report occasional pockets even richer. The mine's website, www.phoenixgoldmine.com, carefully details the original Phoenix Vein, the multiple owners in the early years, and more.

The owners bill themselves as the oldest continuous gold-mining family in Colorado. Just try to stump them with a question! Al gives a great tour, and I can't add much to this testimonial left on his website by a professional oil geologist: "This tour has something most tours lack, the human side of mining. Al captures such an exciting spirit when he gives his tours. The history, mixed with personal anecdotes, and the mining technology he gives on the tour allow you to see the real ups and downs of a time almost lost."

Once back into the sunlight, the panning station is a treat. It's real gold, you keep what you find, and you should do well. Stay as long as you like, work the creek, and have fun!

65. West Chicago Creek

Land type: Creek
County: Clear Creek
Elevation: 8,848 feet at Site A
GPS: A - West Chicago Creek pullout: 39.69314, -105.61653
B - Soda Creek Road: 39.71479, -105.55454
Best season: Summer through fall
Land manager: Arapaho National Forest
Material: Fine gold, small flakes
Tools: Pan, sluice
Vehicle: Any for Site A; 4WD suggested for Site B
Special attractions: Indian Hot Springs; Echo Lake; Underhill Museum in Idaho Springs
Accommodations: Full services in Idaho Springs. Thousand Trails campground on West Chicago Creek Road. Multiple developed campgrounds near here on Chicago Creek. Lodge at Echo Lake. Primitive camping at Hells Hole Trailhead and above Echo Lake.
Finding the site: From Miner Street and 13th Avenue in Idaho Springs, turn south onto 13th and cross I-70 onto Chicago Creek Road/CO 103, which is paved. Drive 6.6 miles and park safely at the first big hairpin, where you'll see a sign for West Chicago Creek Road, which is unpaved. Drive a little farther around the curve in the highway and exit to the right. This is Site A. To reach Site B, backtrack to Idaho Springs and the intersection of 13th and Miner. Turn right onto Miner Street and drive 0.4 mile east, then turn right onto Soda Creek Road. Follow it up for 3.4 miles to the parking area at the end of the road.

Prospecting

Because the Idaho Springs District and surrounding mines are mostly on private land, there isn't a lot of access to good panning locales in this general area. US Forest Service land doesn't start until you get quite a ways up the mountains on this and other drainages. Many of the waterways were straightened and rehabilitated, such as Clear Creek in downtown Idaho Springs, putting them off-limits, too. Plus there is a lot of rafting traffic.

Many of the locals directed us to the first switchback on Squaw Pass Road/CO 103, and we did get some colors around the boulders on the main

Working near the big granite boulders in a small tributary of West Chicago Creek.

creek. Don't work much past the little footbridge, as there is private property to avoid. The parking area is easy to find and has room for several vehicles. Some giant granite boulders serve as traps, and as they decompose, they supply loads of black sands to deal with. There aren't many mines or prospects in the watershed above this location, but it's a pleasant spot with easy access.

The end of Soda Creek Road at Site B has a good parking area with easy access to the Barbour Fork of Soda Creek, but work your way upstream. There is a lot of private land to dodge on both of these roads, so use caution if you are tempted to explore.

66. Argo Mill and Tunnel Tour

Land type: Mine tour
County: Clear Creek
Elevation: 7,508 feet at mine
GPS: 39.74239, -105.50609
Best season: Open year-round; closed Tues
Land manager: Private
Material: Photographs; small flakes at panning area plus an optional gemstone opportunity
Tools: Camera; pan provided
Vehicle: Any
Special attractions: Indian Hot Springs. Clear Creek Supply in Idaho Springs (2448 Colorado Blvd.) has prospecting supplies.
Accommodations: Full services in Idaho Springs. Developed campgrounds nearby along CO 103, plus Columbine Campground outside of Central City. RV parking at multiple spots, such as Cottonwood RV and mobile park.
Finding the site: From Golden, you can drive up US 6 through Clear Creek Canyon, about a 19-mile drive, or take I-70, which is faster and is about 23.5 miles. From I-70, take exit 241, drive about 0.2 mile to the traffic circle, and exit onto Colorado Boulevard. Drive 0.1 mile, then turn right onto Gilson Avenue. Drive less than 0.1 mile, then turn left onto Riverside Drive and go 0.4 mile to the parking area. There is a lot of ongoing construction here, so be flexible.

Prospecting

The Argo Mill dominates the town of Idaho Springs, with its deep red complex of buildings on the north side of Clear Creek easy to spot from the interstate. Following the discoveries of rich gold in 1858 by Tom Golden, John Gregory, and George Jackson, miners had filed about 13,000 mining claims by 1861. As of 1902 some 300 lode mines operated near here, especially to the north on Seaton Mountain between Idaho Springs and Central City. As miners reached greater depths, it became difficult to raise the ore to the surface, and water was a constant issue. Pumps struggled to keep up with the surging groundwater, so finally financiers devised a tunnel system. First proposed by Silas and Ralph Knowles, the tunnel would stretch to Central City, which was

Walking along the family-friendly Argo tour to the mill.

Somehow, I didn't see this sign . . . some of the previous panners weren't very efficient.

2,000 feet higher in elevation, and drain "The Richest Square Mile on Earth," aiding mines at Virginia Canyon, Gilpin Gulch, Russell Gulch, Quartz Hill, Nevadaville, and Central City.

Started in 1893 with hand tools and black blasting powder, the tunnel would be the longest in the world at that time and the biggest mining project to date. It was completed in 1910, and by 1914 a state-of-the-art mill at the end of the tunnel processed ore at full capacity, using ammonium salts, amalgamation, flotation, a cyanide process, and a gravity system such as spirals and Wilfrey (shaker) tables. The mill processed over $100 million of gold ore—around 5 million ounces, worth about $6.5 billion at $1,300 per ounce.

The mine closed in 1943 after four miners died when a dynamite charge deep under the mountain unleashed a torrent of water through the mine, at an estimated power of 500 pounds of pressure per square inch. A piece of equipment bolted out of the tunnel entrance and impaled the hillside across Clear Creek where the Safeway now stands. In 1976 the mine was purchased by a local investor, refurbished, and cleaned up, and soon tours started.

Tours begin hourly from 11 a.m. to 4 p.m., last about an hour, and involve a short bus ride to the top of the property followed by a walk through various parts of the mill. There are lots of mining artifacts, ore specimens, exhibits, photographs, equipment, and information stops, and it's very kid-friendly. The panning troughs include either a gold bag or a gem bag to work through, and knowledgeable experts are on hand to help you get your technique down. Check the website at www.historicargotours.com for additional information about clothing, strollers, weather, etc. Like most mine tours, you don't want to wear flip-flops or sandals, and the temperatures can fluctuate wildly. No backpacks are allowed. During the winter tours, outdoor temperatures are a challenge.

67. Hidee Mine Tour

Land type: Creeks, rivers
County: Gilpin
Elevation: 8,642 feet
GPS: 39.78712, -105.50044
Best season: Open year-round, but winter hours are shortened.
Land manager: Private
Material: Photographs; ore samples from the vein
Tools: Camera; pan provided
Vehicle: Any
Special attractions: Central City casinos; Gilpin History Museum in Central City
Accommodations: Full services in Idaho Springs and Central City. Columbine Campground out of Central City is excellent.
Finding the site: From I-70 below Idaho Springs, take exit 243 for Hidden Valley and Central City, and drive north on Central City Parkway for 6.4 miles. From Central City, drive south on the parkway for 1.8 miles. Turn onto Hidee Mine Road and travel about 0.2 mile to the mine's parking area.

Prospecting

In May 1859 prospector John H. Gregory was working his way through the Clear Creek drainage from his home base in Arapahoe City. He found a rich, heavily oxidized and easily worked gold-bearing ledge at what became the Gregory diggings near present-day Black Hawk. In June of that year, W. Green Russell found heavy gold placers and decomposed lodes in Russell Gulch; by July, there were about a hundred sluices working the area, according to Koschmann and Bergendahl (1968, p. 100). It was reported that some miners earned $400 per day, and $100 per day was common.

The diggings were quickly organized, and miners set to work on the easily processed surface ores that were oxidized. Once they reached the lower ores, which were combined with base metals and difficult to process, production waned. Some ten-stamp mills reportedly recovered only 25 percent of the gold, and none of the silver, zinc, or lead. In 1868 a newer smelter began better recovery, and by 1872 the railroad reached Black Hawk. In all, Gilpin County is estimated to have produced 4.2 million ounces of gold, with only

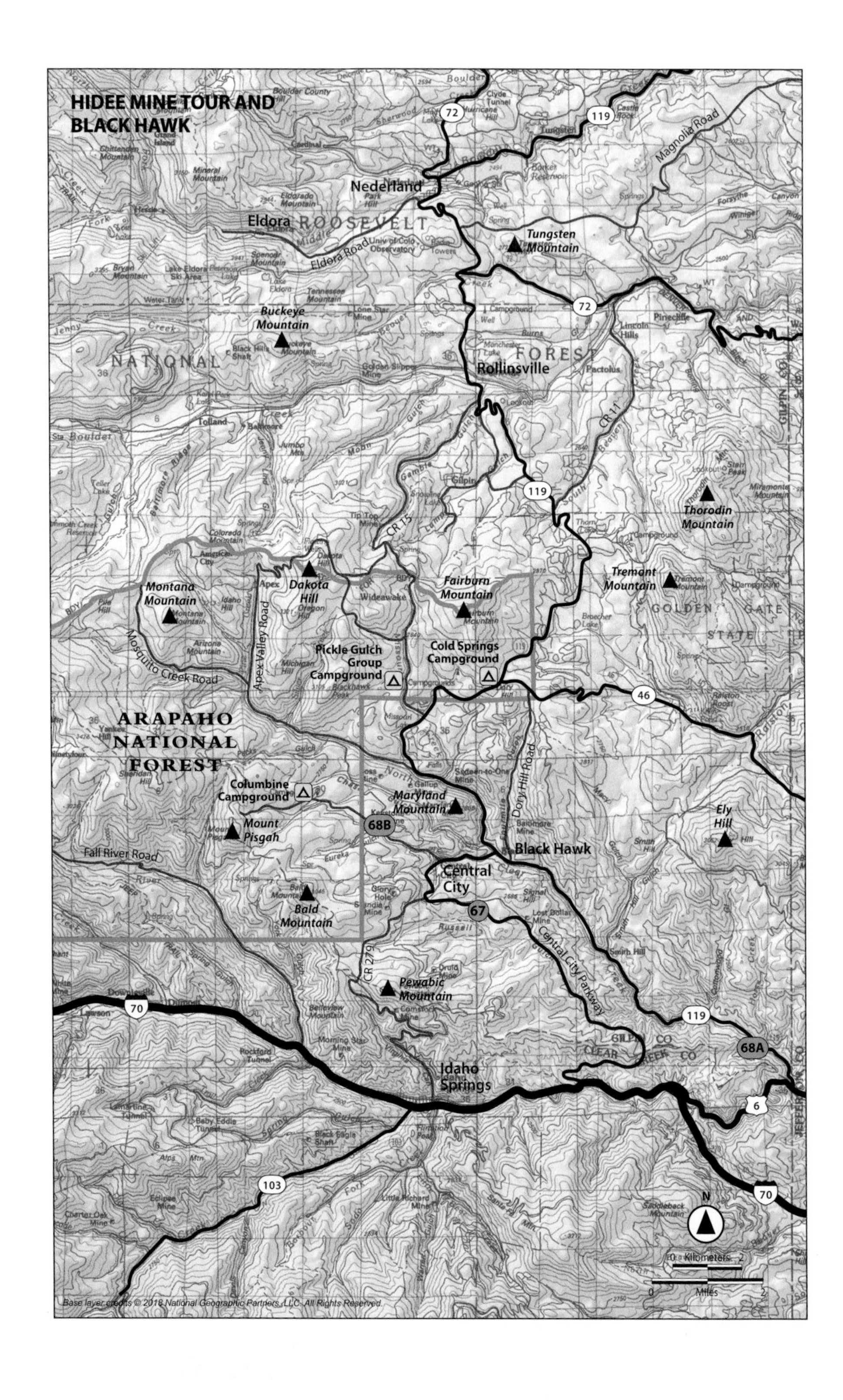
HIDEE MINE TOUR AND
BLACK HAWK
Nederland
Eldora
ROOSEVELT
NATIONAL
FOREST
Tungsten Mountain
Buckeye Mountain
Rollinsville
Eldora Road
Magnolia Road
Thorodin Mountain
Tremont Mountain
GOLDEN GATE
STATE
Montana Mountain
Dakota Hill
Fairburn Mountain
Apex Valley Road
Mosquito Creek Road
Pickle Gulch
Group
Campground
Cold Springs
Campground
ARAPAHO
NATIONAL
FOREST
Columbine
Campground
Maryland Mountain
Mount Pisgah
Fall River Road
Black Hawk
Central
City
Dory Hill Road
Bald Mountain
Ely Hill
Central City Parkway
Pewabic Mountain
Idaho
Springs
CR 15
CR 11
CR 279
72
119
46
67
68B
68A
70
6
103
N
0 Kilometers 2
0 Miles 2
Base layer credits © 2018 National Geographic Partners, LLC. All Rights Reserved.

Inside the well-lit Hidee Mine tour, headed for the ore vein to hammer out our own samples.

about 48,000 from placers. The Central City District produced all but 85,000 ounces of that total; the North Gilpin District was much less important.

The Hidee Mine sits in the heart of the Virginia Canyon–Glory Hole Area, reputed to be the richest square mile on earth. Specimen-quality pyritic gold ore is the principal product. The mines in this area all told produced more than $5 billion (2012 gold value) in gold and silver since 1859. Ore produced from the Hidee Mine has been valued as high as 7.9 ounces gold, 14.6 ounces silver, and 9.6 percent copper per ton. Assays of mine samples have been as high as 112 ounces gold, 20 ounces silver, and 16.5 percent copper per ton.

The geology is very similar to the Clear Creek District; Koschmann and Bergendahl (1968, p. 93) treat them as the same. "The country rock . . . is a complex of Precambrian metamorphic and igneous rocks cut by Tertiary stocks and numerous dikes of porphyries. The most common porphyries range in composition from quartz monzonite to bostonite and alaskite. The ore deposits are also Tertiary in age. The ore bodies are in pyritic quartz veins with gold and copper; chief minerals are pyrite, chalcopyrite, and quartz, with sphalerite and galena. The gold is associated with the chalcopyrite."

The Hidee Mine tour is much more intimate than most. It's also famous in its own right; check www.hideegoldmine.com for the numerous testimonials and write-ups, ranging from *USA Today* to CBS TV. Being a bit off the beaten track, tour groups can be small. Visitors walk into the cool underground workings along a level floor, then descend to a lower area where the operator will introduce you to the pyrite-rich vein. The operators have a collection of hammers and chisels to attack the face, and you'll receive a small cloth bag to cram full of samples. Outside at the panning trough, which costs extra but is still modest, you will learn to pan the mine's pyrite ores, with a few gemstones tossed in.

68. Black Hawk

Land type: Creek
County: Gilpin
Elevation: 7,066 feet at Site A; 8,916 feet at Site B
GPS: A - Emergency phone spot on North Fork Clear Creek: 39.75537, -105.40939
B - Boodle Mill: 39.80789, -105.53036
Best season: Late spring through fall
Land manager: BLM
Material: Fine gold, small flakes
Tools: Pan
Vehicle: Any
Special attractions: Central City casinos; Gilpin History Museum in Central City.
Accommodations: None at site; Idaho City has full services. Columbine Campground is the closest developed campground; Cold Springs Campground is also nearby. Pickle Gulch Campground is for groups only.
Finding the site: From the intersection of Black Hawk Street and CO 119, drive southeast on CO 119 for 6.3 miles to reach Site A. Look for an emergency phone located in a wide gravel area on the right, about 0.4 mile past the gas station. To reach Site B, drive back to Black Hawk and turn left onto Black Hawk Street, which becomes Gregory Street after 0.5 mile. Drive another 0.6 mile and the main road becomes Lawrence Street in the old part of town. Stay on the main road and continue straight on Eureka Street for 0.8 mile, where the road now becomes Upper Apex Road. Drive 0.4 mile and the mine is right there.

Prospecting

Use caution with Site A and be flexible. I followed advice from "Red Stater" at www.goldstrikecoloradoprospecting.blogspot.com, but it's dated to 2011. Kevin Singel had a long thread at www.findinggoldincolorado.com about North Clear Creek, but he eventually ended the discussion because there were questions about access. Here's the listing I found, lightly edited:

> *North Fork of Clear Creek—Highway 119 to Black Hawk. You can get access just downstream from the convenience store which is about 2 miles upstream of the split from Highway 6. There are some holes already started which will deliver some fine gold in spots. Trying to work in the streambed on*

The Boodle Mill is an easy landmark near Columbine Campground and the Central City cemeteries.

> *North Fork is futile at best. The soil is heavy clay material with tons of silt from the old mine tailings. There is also a small piece of BLM land marked by a roadside emergency phone upstream about 2 miles below Black Hawk which has produced some fine gold from the banks and benches up above the stream itself.*

When we visited, there weren't any postings at the two sites, probably because of the cautions above and the fact that the open public area on Clear Creek is just a few miles away. Still, the idea of being able to pan just below Russell Gulch was hard to pass up, so I kept the site at the emergency phone "just in case."

Nearby, you can also try Boulder Creek out of Rollinsville, at 39.91481, -105.53768, for some colors. We scouted this one out and didn't see any postings, and there were several mines out here. I was hoping to camp there and use it as a base station for more recon, but the camping area has been converted to an interpretive area. We took the backcountry route from here all the way to the west end of Central City, passing the many cemeteries near the Boodle Mill. There was talk of rehabilitating it for mine tours, so keep an eye out for that.

Also, back in Central City, the Coeur d'Alene Mine conducts tours of the shaft house. Nearby, south of Central City, try the road to Nevadaville, a picturesque little ghost town. Panning lessons used to given there, conducted by the two residents. Upper Russell Gulch Road, via CR 279/Virginia Canyon Road, will also take you to some interesting ruins. If you go farther west to Montana Creek and Mosquito Creek, you'll be on public land again and you can do more scouting, or try north up near Jamestown. There are a lot of opportunities left to check up here at the northeast end of the Colorado Mineral Belt.

69. Clear Creek Public Area

Land type: Mountain creek
County: Jefferson
Elevation: 6,854 feet at Site H
GPS: A - East end/Tunnel 1: 39.74770, -105.25111
B - Rapids: 39.73876, -105.25988
C - Big parking area: 39.73869, -105.26682
D - Best beach: 39.74183, -105.28605
E - Tunnel 2: 39.73967, -105.32268
F - Private claim east end: 39.74159, -105.33164
G - Private claim west end: 39.74073, -105.35167
H - West end: 39.74229, -105.39063
Best season: Late spring through fall
Land manager: Jefferson County Open Space
Material: Fine gold, small flakes; bigger flakes and small nuggets possible
Tools: Pan, sluice, dredge (with permit)
Vehicle: Any
Special attractions: Idaho Springs
Accommodations: Full services in Golden; Columbine Campground is the closest developed campground; Cold Springs Campground is also nearby. Pickle Gulch Campground is for groups only.
Finding the site: From downtown Golden, drive northwest on Washington Avenue to CO 58. Turn left and merge onto CO 58 going west. Drive 0.7 mile west, just past the intersection with CO 93, where CO 58 now becomes US 6/Clear Creek Canyon Road. Continue westbound on US 6 for 1.1 miles. The parking area on the left marks the eastern end of the free public area. To reach Site B, continue west on US 6 for about 0.9 mile to the parking area on the left. To reach Site C, drive another 0.4 mile west on US 6 and note the large parking area on the left. To reach Site D, drive farther west on US 6 for about 1.4 miles. This is probably the best beach in the canyon and there is ample parking, so it could be busy. To reach Site E, which is the west end of Tunnel 2, drive about 2.8 miles. The parking on the left is a lot easier to reach from the eastbound lane than Tunnel 3. To reach Site F, drive another 0.5 mile west, past the parking area for the west end of Tunnel 3. The tunnel parking spot is almost impossible to hit from the eastbound lane, so I didn't list it. The private claim starts just west of here and stretches about 1 mile west to

This is the best beach area in the canyon, marked as Site D. It's been hit hard, but still produces.

Site G. There is a parking area behind the guardrail; you can park here and work the area—just stay upstream from the east end of the guardrail. To reach Site H, which is the farthest western spot on Clear Creek that is open to the public, drive on US 6 west for 2.8 miles. You'll see a big sign here marking the start of public panning.

Prospecting

Tom Golden, John Gregory, and George Jackson all worked their way up from the strikes at Arapahoe Bar and Cherry Creek to explore the Clear Creek drainage in 1859. They undoubtedly found colors just like you'll find, enough to convince them that there were good lode deposits higher up. Jackson found lode gold at what is now Idaho Springs in January 1859, and Gregory located a nice ledge near what is now Black Hawk in May 1859. Their persistence and hardships paved the way for the great "59er" gold rush into Colorado.

Parker (2009, pp. 39–41) describes many of the best places to check above the public area, including terraces far above the current creek. You'll find several YouTube videos about dredging, chasing culvert gold, and more for

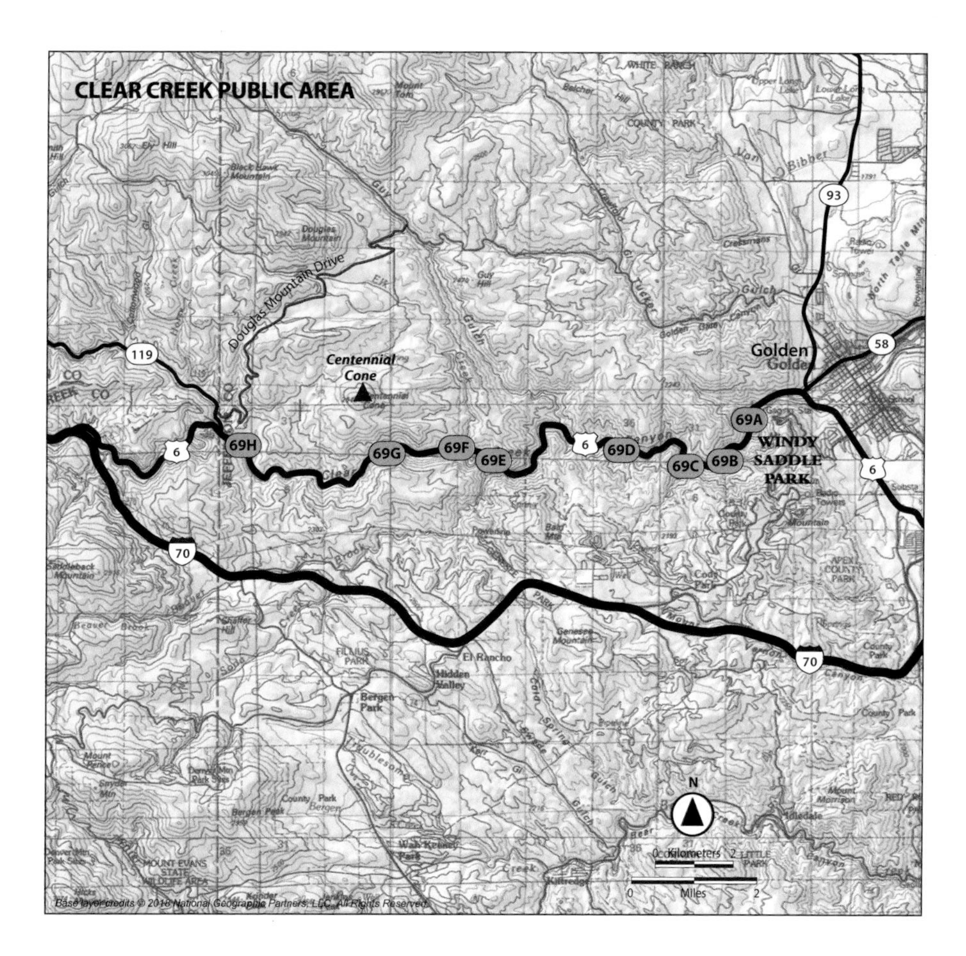

this park. This is undoubtedly the premier public panning spot in the northeast sector of the Colorado Mineral Belt. It runs roughly from mile marker 260.5 to 270.5, with the exception of a single private claim near the middle, according to prospecting sites. The water runs fast and clean, the gold is good, dredging is permitted, and there are plenty of access points. You can use your battery-powered equipment here, too, which is a real plus. You'll need a permit to run gas-powered machines.

I listed the coordinates for the only private claim in the corridor so you can steer clear of it. Everything else is open.

Naturally, there are plenty of rules. Two in particular stand out:

1. You can't leave your equipment in the canyon overnight and expect to come back for it. That is considered abandoning property, akin to littering, and will net you a big fine. Come early, leave late, and pack up your stuff.

2. Do not work within 100 feet of structures such as bridges, steps, and fish habitat improvements. It took a lot of work to create those things, and you'll get ticketed for messing with them.

At Site A, park safely and work your way upstream along the footpath. Ideally, you can get to the upper end of the giant bend here at Tunnel 1, where there are small beaches. It's a long way to hike in your equipment, and a long way to hike back buckets, but it's a good place to start your Clear Creek Canyon prospecting.

Site B is pretty good when the water is low—there are rapids downstream and big rocks right at the parking area. Site C has even more parking, with big rocks upstream. There is an excellent beach at Site D, and the upstream end has good colors. One prospector I met up here said he was surprised to pull out 2 grams, as he expected it to be hit hard, being so easy to reach.

Site E is the western end of Tunnel 2 and has ample parking. There isn't a lot of good trail here, however; it's a scramble to reach the best part of the bend that the tunnel cuts through. Sites F and G mark the 1-mile stretch of private land to watch out for. There are no claim markers up that I saw, but I didn't see any spots that looked any better than the rest of the canyon, so it's no big loss if it's still privately claimed. I got my information from www.goldcube.net/2016/05/09/clear-creek-canyon-park, so if conditions change, that website might know. Site H is the western end of the public area.

70. Gold Strike Park Public Area

Land type: Creek in city park
County: Jefferson
Elevation: 5,271 feet
GPS: 39.79798, -105.05729
Best season: Late spring through fall
Land manager: City of Arvada
Material: Fine gold, small flakes
Tools: Pan, sluice; nonmotorized equipment OK
Vehicle: Any
Special attractions: Denver museums; Coors Brewery in Golden
Accommodations: Full services in Golden and Denver. No camping in public area. Columbine Campground is the closest developed campground; Cold Springs Campground is also nearby. Pickle Gulch Campground is for groups only.
Finding the site: From Golden, drive east on I-70 about 7 miles, then take exit 269B on the left to connect to I-76. Drive 1.3 miles on I-76, then take exit 1B for CO 95/Sheridan Boulevard. Drive 0.3 mile through the exit ramp to turn left onto CO 95/Sheridan Boulevard. Drive just 0.3 mile and turn left onto Ralston Road, then drive 0.2 mile and turn left again onto W. 56th Avenue. After 0.1 mile you'll see the entrance to Gold Strike Park on your left. The coordinates are from the parking lot.

Prospecting

Lewis Ralston was already a veteran gold prospector when he arrived in the Colorado Territory in 1850. A Scotsman who married the granddaughter of a Cherokee chief, he had previously lived in Georgia when gold was discovered near his land near Auraria in 1828. After getting a taste of gold fever, and hearing about the excitement at Sutter's Mill in California, Ralston joined an expedition headed west, captained by Clement Vann McNair, in 1850. After fording the South Platte River near its confluence with Clear Creek, the party rested on an unnamed creek, and it was there that Ralston found about $5 in gold in his first pan. The party named the creek for him, but they did not stop to work it, as they were intent on reaching California.

Most of that party failed in California and returned home within a year, but in 1858 Ralston agreed to guide a party led by William Greeneberry "Green" Russell back to that creek in Colorado. About 100 men arrived in

View upstream on Ralston Creek from under the highway bridge. Please don't dig around the bridge's concrete foundations!

late June, but most had left by July 4. Those that remained discovered a rich pocket of gold on Little Dry Creek, and by November they had founded the town of Auraria at the mouth of Cherry Creek. Denver City was established nearby, and in 1860 the two towns merged.

The pedestrian bridge here was built in 2001, and the site is open for recreational, nonmotorized prospecting on both Ralston and Clear Creeks. That doesn't seem to include battery-powered contraptions, but maybe someone can force the issue and get that rule changed. Running a Gold Cube or a spiral pan in a tub shouldn't be a problem, but it seems to be a gray area here.

One advantage at Ralston Creek is you can work under the highway bridge (much bigger shadow than the pedestrian bridge, which crosses Clear Creek) and get some shade during the summer. Or you can follow Ralston Creek down to its mouth on Clear Creek, either by taking a path or just walking the creek. You'll be able to confirm that Ralston Creek runs cleaner than Clear Creek, by the way.

Whenever working at the mouth of a creek, I like to work on the downstream edge of the mouth on the bigger creek, as this area will always trap gold during high-water flows. If you live in the area, it's always a good idea to check the creek at different times of the year and try to get an idea of what the water is doing when it is at its highest flood, so that when you come back when the water is low, you know where to sample and then set up a sluice.

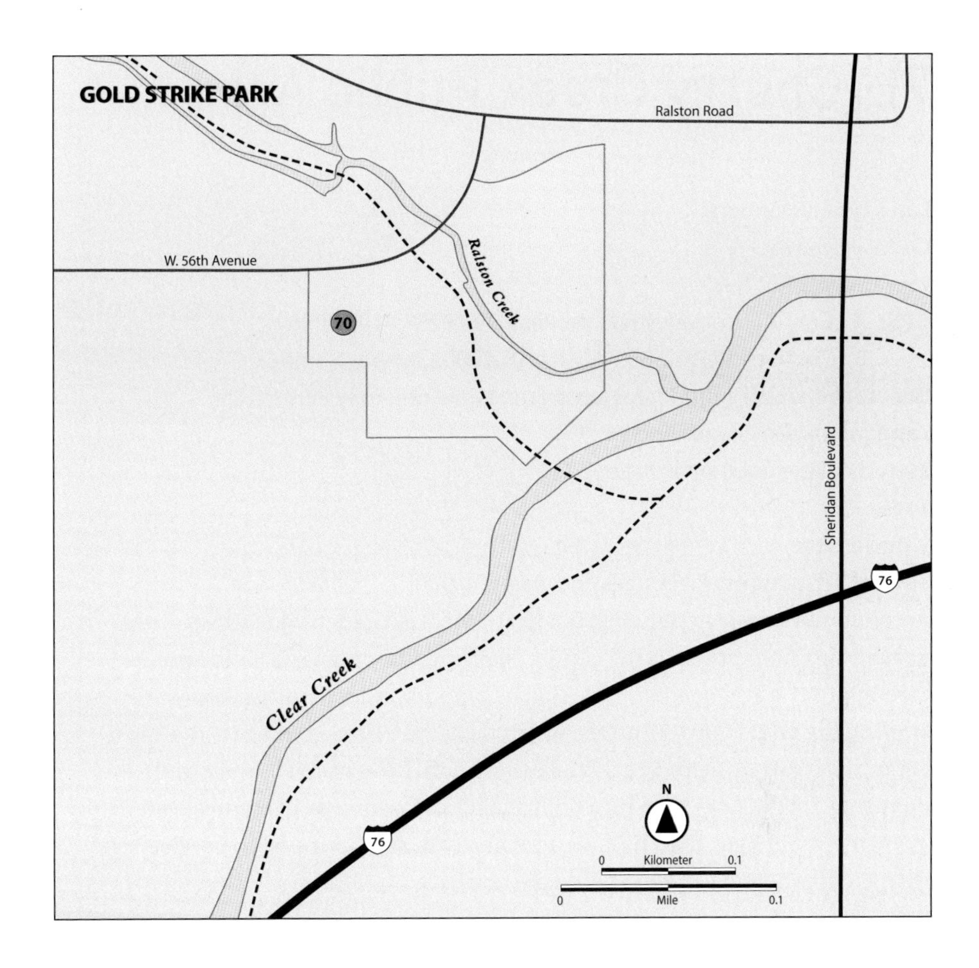

Remember to be a good citizen here and bring out some trash to help clean up the area. Also, even though you might see where some prospectors have dug against the bridge foundations, please don't. A highway engineer would completely freak out if they saw you doing that, and it could lead to the area being shut down. Stick to the creek and the immediate banks—but don't dig up vegetation.

Both "Red Stater" at www.goldstrikecoloradoprospecting.blogspot.com and Kevin Singel at www.findinggoldincolorado.com have write-ups of this area. "Red" also mentions Memorial Park and Hoskinson Park as having access, although he warns about broken glass and lots of trash. Singel reports that he did well in the first 6 inches of gravel, which we confirmed, but another prospector reported good pieces of gold 2 feet down.

71. Cherry Creek Public Area

Land type: Riverbank
County: Denver
Elevation: 5,174 feet
GPS: A - Confluence Park street parking: 39.75267, -105.00869
B - Confluence Park: 39.75471, -105.00802
Best season: Late summer for low water; midwinter for cleaner water
Land manager: City of Denver
Material: Fine gold, small flakes
Tools: Pan
Vehicle: Any
Special attractions: River trails and walks
Accommodations: No camping, but full services in Golden and Denver. Columbine Campground is the closest developed campground; Cold Springs Campground is also nearby. Pickle Gulch Campground is for groups only.
Finding the site: From I-25 on the west side of the Platte as you pass the Pepsi Center, take exit 212A for Speer Boulevard South. Drive about 0.2 mile and turn right onto Elitch Circle, then turn right again to get closer to Centennial Park. Park along the street here, then hike west to the Platte and northeast to Confluence Park. Cross over and work the bar to the north.

Prospecting

Cherry Creek offers a taste of what can be called "urban gold panning." These spots are along the lower part of Cherry Creek, where it merges with the South Platte River at Centennial Park. It doesn't get much more urban than this—the Pepsi Center, where the Denver Nuggets play basketball; Coors Field, where the Colorado Rockies play baseball; and Mile High Stadium, where the Denver Broncos play football, are all nearby. Thus, parking can be an issue. So can street people, gawkers, joggers, cyclists, and skateboarders. Plus, the water can be a bit murky by summer; if you can brave the cold, you might want to try in the winter, when the water is super low and the bacteria count is lowest.

From its mouth, the Cherry Creek drainage stretches south through Denver and Glendale to Cherry Creek Lake, then farther south past Franktown to Castlewood Canyon. The Cherry Creek Trail offers some access to the upper

For urban panning at its finest, try Cherry Creek where it enters the South Fork of the Platte at Confluence Park.

locales, such as at Cottonwood Park, but I didn't try it. These placers were abandoned long ago, and development has overtaken good spots in the upper drainage. If you know someone who lives along Russellville Gulch, you might have a chance. Everything else looks gated, developed, or otherwise off-limits.

Kevin Singel's fine website at www.findinggoldincolorado.com has more good information; Kevin haunts this area regularly and has pulled quite a bit of gold from the creek. In addition to this area, he's sampled just upstream of the Colorado Boulevard bridge over the creek in north Glendale. At www.goldstrikecoloradoprospecting.com, Cherry Creek is listed as one of the more challenging locales, as it is very fine gold. Here's part of the report there: "The first worthwhile gold strike was on the South Platte River, in Denver County, near Overland Park between 8th Ave and Jewell. The most productive deposits found were on Big Dry Creek, Newlin and Russellville Gulches in the Cherry Creek drainage in Douglas County, and on Ronk (now called Gold Creek) and Gold Run gulches to the east of Russellville gulch in Elbert County."

More good news and bad news: The gold is exceptionally pure in the Cherry Creek drainage—as high as .990 fine. But at 50 colors per milligram, this is essentially "flour gold," or "fly-poop" as one miner described it. The source is apparently the Castle Rock Conglomerate, which houses ancient streambeds ranging from sand and silt to pebbles and cobbles, held together in a clay matrix. The conglomerate is quite erratic, but Cherry Creek concentrated the heavies in its many twists and bends. Some sources indicate you

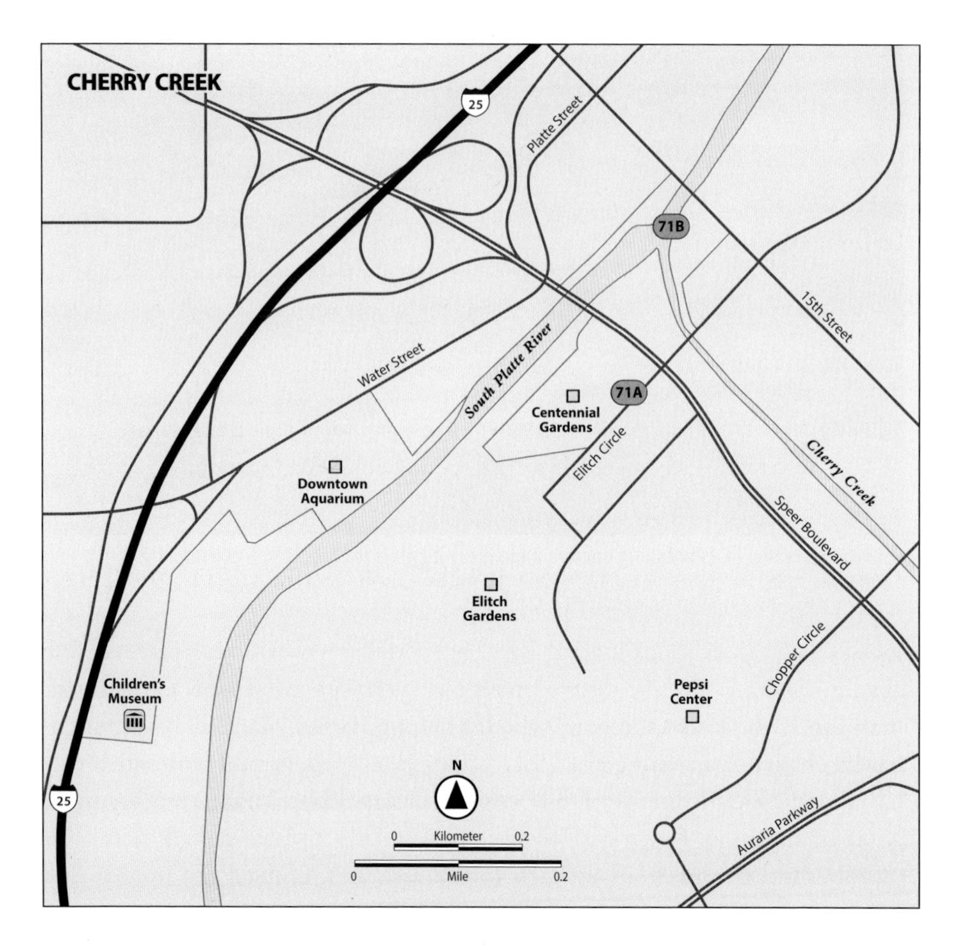

could find petrified wood or even fossil mammal bones in the creek, so be on the lookout. Also, heavy storms can bring the water level up quickly if the Cherry Creek Dam needs to reduce its contents, so beware of that hazard.

Barring all that, try the mouth of Cherry Creek. The Outdoor Channel features a show called *Gold Fever* with the enthusiastic Tom Massie, and in one episode, he and his daughter used a hand-powered suction device to slurp up gravels at Site B. They were working the lip of a man-made flow constrictor on the Platte and managed to find a nice flake. The city has put so much work into landscaping the area and making it look nice that you don't want to disturb any vegetation or dig in the banks, but the sandy beach should yield some colors. Be sure to fill in your holes and maybe pick up a piece or two of garbage while you're there, just to show your heart is in the right place.

72. Arapahoe Bar Public Area

Land type: City creek
County: Jefferson
Elevation: 5,434 feet
GPS: A - Arapahoe Bar parking: 39.77322, -105.14225
B - Clear Creek walk-in: 39.77201, -105.14382
C - Arapahoe Bar: 39.77337, -105.14542
Best season: Late summer for low water; winter months for fresher water
Land manager: City of Wheat Ridge Parks and Recreation Department (www.ci.wheatridge.co.us/1184/Gold-Panning)
Material: Fine gold, small flakes
Tools: Pan, sluice; non-gas-powered equipment only. Devise a battery box to keep your battery out of the water.
Vehicle: Any
Special attractions: Coors Brewery in Golden
Accommodations: No camping at site; full services in Golden
Finding the site: The area may be accessed by parking at either the 41st and Youngfield trailhead parking lot or on the south frontage road of CO 58, east of McIntyre Street.

Prospecting

The first explorers to discover gold on Clear Creek in this vicinity were part of the Estes Party of 1834, but few records of their discovery exist. In 1858 the Doniphan Party, including Marshall Cook, rediscovered the old workings and laid out new claims. They founded Arapahoe City in November 1858, and soon there was a complete town with a post office and wooden buildings. Two early citizens of Arapahoe City stand out: John Hamilton Gregory, who arrived in 1859, and George Andrews Jackson, who arrived in December 1958. Jackson and his partner, Thomas L. Golden, explored up the watershed. All alone, Jackson found the lode gold at what is now Idaho Springs in January 1859. Gregory located one of the biggest veins near what is now Black Hawk in Gilpin County in May 1859. The prospecting by Jackson and Gregory sparked the great "59er" gold rush into Colorado.

By the late 1860s, not much was left of Arapahoe City, as the easy gold was exhausted via hydraulics fed by ditches and also by dredging. Many of

Clear Creek is easy to wade across to reach Arapahoe Bar. Stay clear of the south bank here.

the buildings were moved to Golden, and today only a stone marker remains at 39.77511, -105.17826, about 1.75 miles to the west, near one of the Rio Grande pits.

Recreational gold panning is allowed in Clear Creek at the Arapahoe Bar area without fees or permits. There are two access points: the small parking area on the west side of Youngfield Service Road, and the much larger Wheat Ridge Greenbelt Youngfield Entrance on the east side of Youngfield Street. I recommend the latter spot, as it has a restroom.

Do not work on the south side of Clear Creek, and do not venture east beyond the bridge. No form of gold panning is allowed in Clear Creek east of Youngfield in the Wheat Ridge Greenbelt. Clear Creek can be a bit "spicy"

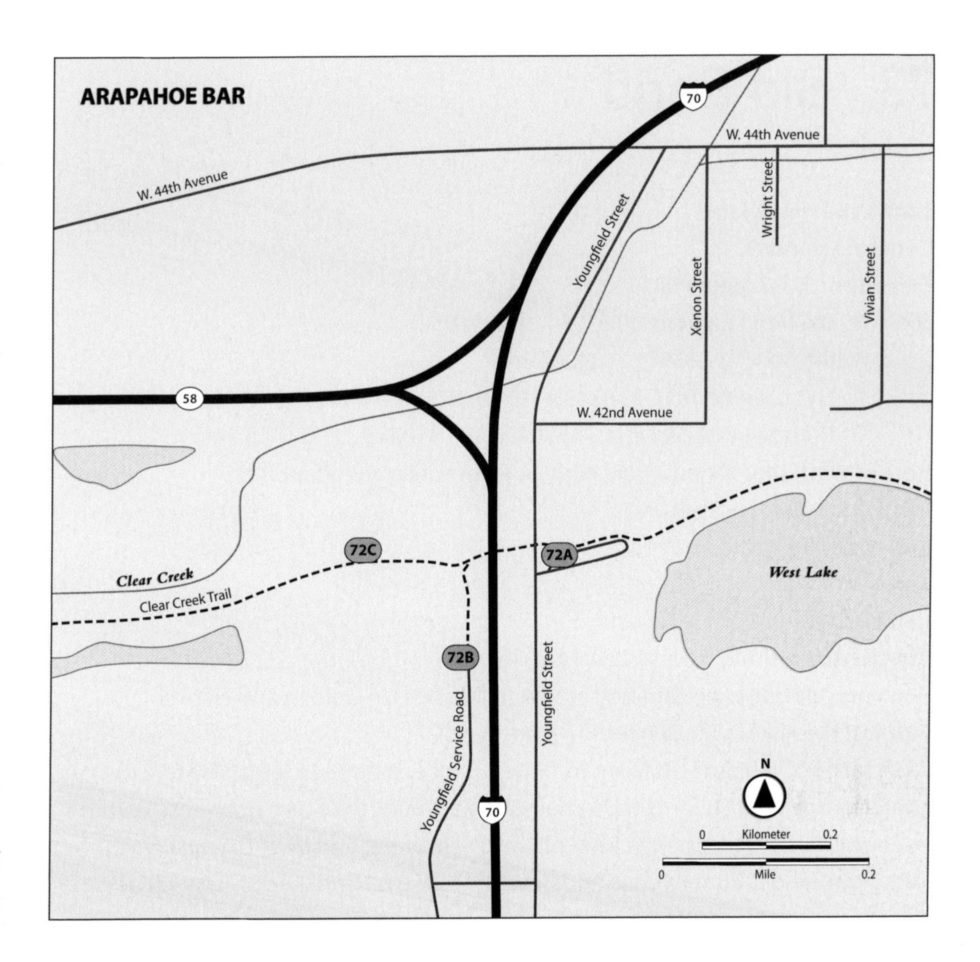

down here below the Coors Brewery, so colder months might be a good idea. The place has been hammered, but that's because there is good gold here, and it keeps washing down during each flood season. Do your best to help the area look good. Fill in your holes, pick up trash, don't disturb the vegetation, stay in the correct area, mind the signs—all that should be second nature anyway. Being in a public area, you want to do your part to keep this location open for future recreational panners.

73. Big Bend

Land type: Riverbank
County: Adams
Elevation: 5,087 feet at Big Bend
GPS: A - Big Bend parking: 39.83934, -104.94956
B - Big Bend: 39.83914, -104.94756
C - Front Range Trail: 39.83435, -104.94871
D - 74th Avenue parking: 39.83049, -104.95031
Best season: Year-round; avoid high water in spring runoff months.
Land manager: Local cities
Material: Fine gold, small flakes
Tools: Pan, sluice
Vehicle: Any
Special attractions: Coors Brewery in Golden
Accommodations: No camping at site; full services in Golden and Denver
Finding the site: From Denver, drive north on I-25 about 4.3 miles, then take exit 216A for I-76E toward Fort Morgan. Drive 0.2 mile, then keep left at the fork to continue to exit 216B for 70th Avenue. Drive 0.6 mile, then turn right onto E. 70th Avenue. Drive 1.3 miles, then turn left onto York Street and drive 0.7 mile north. Turn right onto E. 78th Avenue and drive 0.5 mile east. The road swings left at the river and turns into Steele Street; continue about 0.1 mile and turn right into the parking area. The coordinates for Site A are in the parking area. To reach Site B, walk south down the sidewalk to the junction with the Colorado Front Range Trail and turn left. Walk about 0.1 mile under the railroad bridge, headed east, downriver; you can't miss the big gravel bar. My coordinates are for the top of the bar where the heaviest particles should be settling. To reach Site C, return to the Front Range Trail and walk along the river, past the parking area, and another 0.3 mile to a good beach area when the water is low. To reach Site D from the parking area at Site A, return to Steele Avenue and head south, then west on E. 78th to York. Turn left, drive 0.6 mile, and turn left onto E. 74th Avenue. Drive 0.5 mile east, and look for the turnout to the parking area. The coordinates for Site D are for the parking lot. You could reach Site C by hiking north, down the river about 0.3 mile, or head for the mouth of Clear Creek, about 0.2 mile south, upstream. There is a footbridge across Clear Creek, so you can reach both sides of the creek here.

Prospecting

Koschmann and Bergendahl (1968, p. 87) credit Clear Creek placers in Adams County with producing about 16,800 ounces of mostly fine gold through 1959. There are no breakouts for production on the South Fork of the Platte River, and most of its production, like that of Clear Creek, has probably been incidental to sand and gravel operations starting in 1922. Still, there is fine but consistent gold in Clear Creek and here on the South Platte. Site B is not really Big Bend—that name is more accurately applied to the large bend on the other side of the river, just upstream. If you spend an afternoon here at Site B, try the upper part of the bar. Clean out a hole at least 24 inches deep and see what you find.

Somewhere along here was featured in an episode of *Gold Fever,* when Tom Massie set up a competition between teams formed from the local prospecting club. They worked the area when there was snow on the ground, so don't think of this location as a summer-only option. During the colder months the water can be at its very lowest, and it won't smell too bad either.

It looks like you can reach the actual Big Bend spot from the Fernald Trail. There is a parking area at 39.83085, -104.94653, and it looks like dirt trails go north along the river. I can't vouch for it, though.

You can reach Site C from either the parking area at Site A or the parking at Site D. But by far your best bet is the mouth of Clear Creek, south of the parking area at Site D. There is a large jumble of boulders at the mouth that you could work upstream, and a gravel bar forms there on the north side of Clear Creek. When the water is low, another gravel bar forms upstream on Clear Creek above the footbridge. If you want to mount an expedition up Clear Creek via bicycle, you'll find another good gravel bar upstream from the York Street bridge. There is no good parking in the area, so using a bicycle with your fold-up sluice in a backpack would be a good idea.

Adams County allows recreational gold prospecting using nonmotorized equipment (no dredges and gas-powered high-bankers, according to the county parks department) in all its open space and parks property along Clear Creek and the South Platte River. These are the most obvious coordinates to start with; as you can imagine, anything easy to get to probably gets hammered every summer. Denver and Golden and the surrounding cities are unique with their openness to recreational prospecting inside the city limits. Parking at any of the trailheads and moving far away to find secluded spots might mean better gold.

During low-water months, the South Platte exposes a large gravel bar below the railroad bridge, with fine gold in the cobbles.

Here are some other coordinates to consider, from east to west:

Platte River Trailhead Park: 39.85452, -104.93975. Good access to the river, but getting pretty far below the mouth of Clear Creek.

70th Avenue near Gilpin Way: 39.82636, -104.96721. There is a little parking area on the south side of 70th where it used to be possible to pull in. It may be closed now. Don't be tempted to park in the industrial complex along Gilpin Way—you could easily get towed.

70th and Washington: 39.82410, -104.97758. We thought about stopping in for "recon" tattoos for the crew at the Blue Rose tattoo shop, which would give us permission to park close to the Washington Street bridge locations along Clear Creek. Cooler heads prevailed. If you can get permission to park, that's great, but you still have to schlep your gear through the intersection.

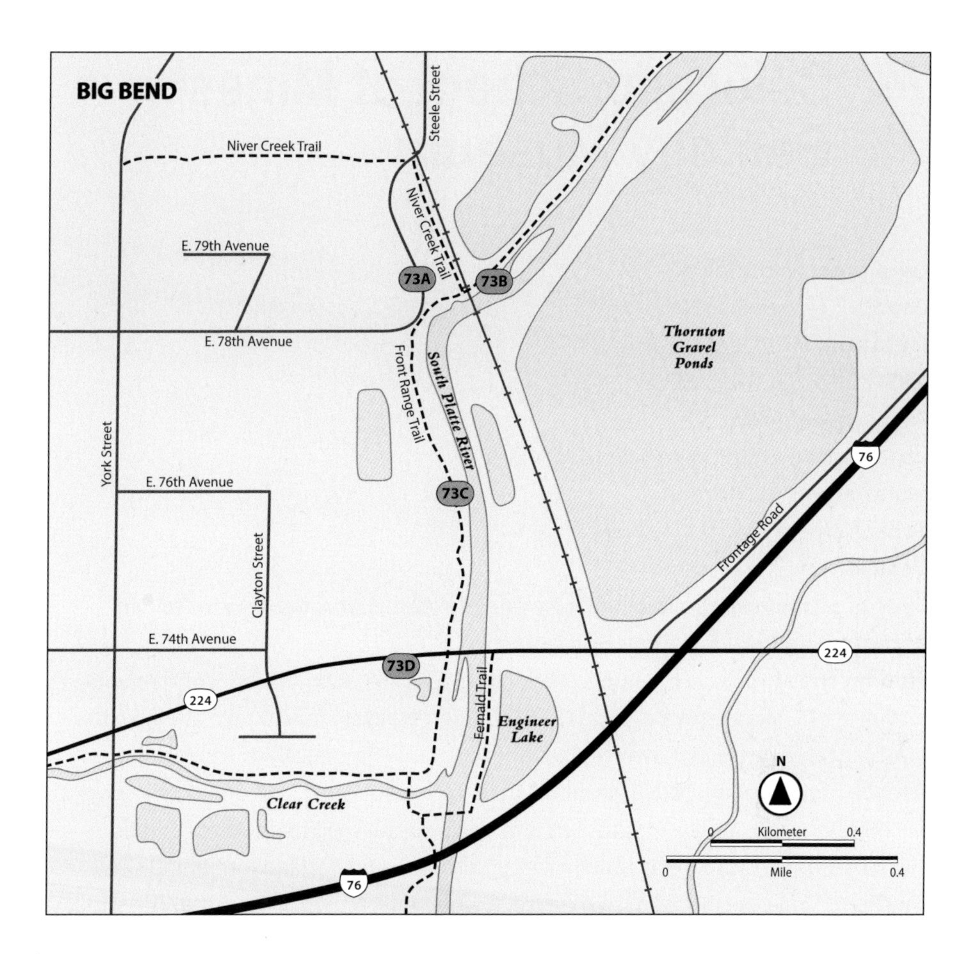

CO 53 bridge: 39.81989, -104.98781. There is a large gravel parking area at the south end of the business park, but it just looked sketchy. With the gate open, you could drive along the Clear Creek Trail, but that didn't seem right. One rig with out-of-state plates was parked here; I would have felt better if they had been local. The gravels under and upstream from the CO 53 bridge looked tantalizing.

Twin Lakes parking: 39.82333, -104.99025. This would require a bicycle assault, as it is 0.6 mile to the CO 53 bridge and 0.75 mile upstream to some promising sandbars.

Lowell Boulevard: 39.79715, -105.03419. Good access to Clear Creek gravels downstream.

74. Colorado School of Mines Geology Museum

Land type: Ghost town
County: Denver
Elevation: 5,790 feet
GPS: 39.75193, -105.22511
Best season: Any
Land manager: Colorado School of Mines
Material: Gold collection
Tools: Camera
Vehicle: Any
Special attractions: Buffalo Bill Grave and Museum; Coors Brewery in Golden
Accommodations: Full services in Golden
Finding the site: From the intersection of US 6/CO 58 and CO 93/Second Avenue just west of Golden, drive east on CO 58/Golden Freeway for 0.6 mile and take the Washington Avenue exit, which takes about 0.1 mile to negotiate. At Washington Avenue, turn right and drive about 0.3 mile. Turn right onto 11th Street and drive 0.3 mile, where 11th Street turns left and becomes Maple Street. After 0.1 mile turn right onto 13th Street, which takes you to the paid parking lot behind the museum. Since the museum tour is free, it won't hurt to pay for parking. The street address is 1310 Maple St.

Prospecting

Started in 1874, the Colorado School of Mines Geology Museum is a must-see for anyone interested in recreational gold panning in the state of Colorado. They have one of the best gold collections you'll ever see, plus moon rocks, fossils, meteorites, mining artifacts, and much more. There are exhibits about radioactivity, fluorescence, paleontology, and underground mining. A walking tour around the school grounds features fossils and dinosaur tracks along the way. The gift shop is also well-stocked and full of treasures.

Thanks to my affiliation with the Rice Northwest Museum of Rocks and Minerals, I got an introduction to the museum director, Dr. Bruce Geller, and he gave me a tour around the facility. He explained the energy and enthusiasm present among volunteers and staff, the great connection to school alumni,

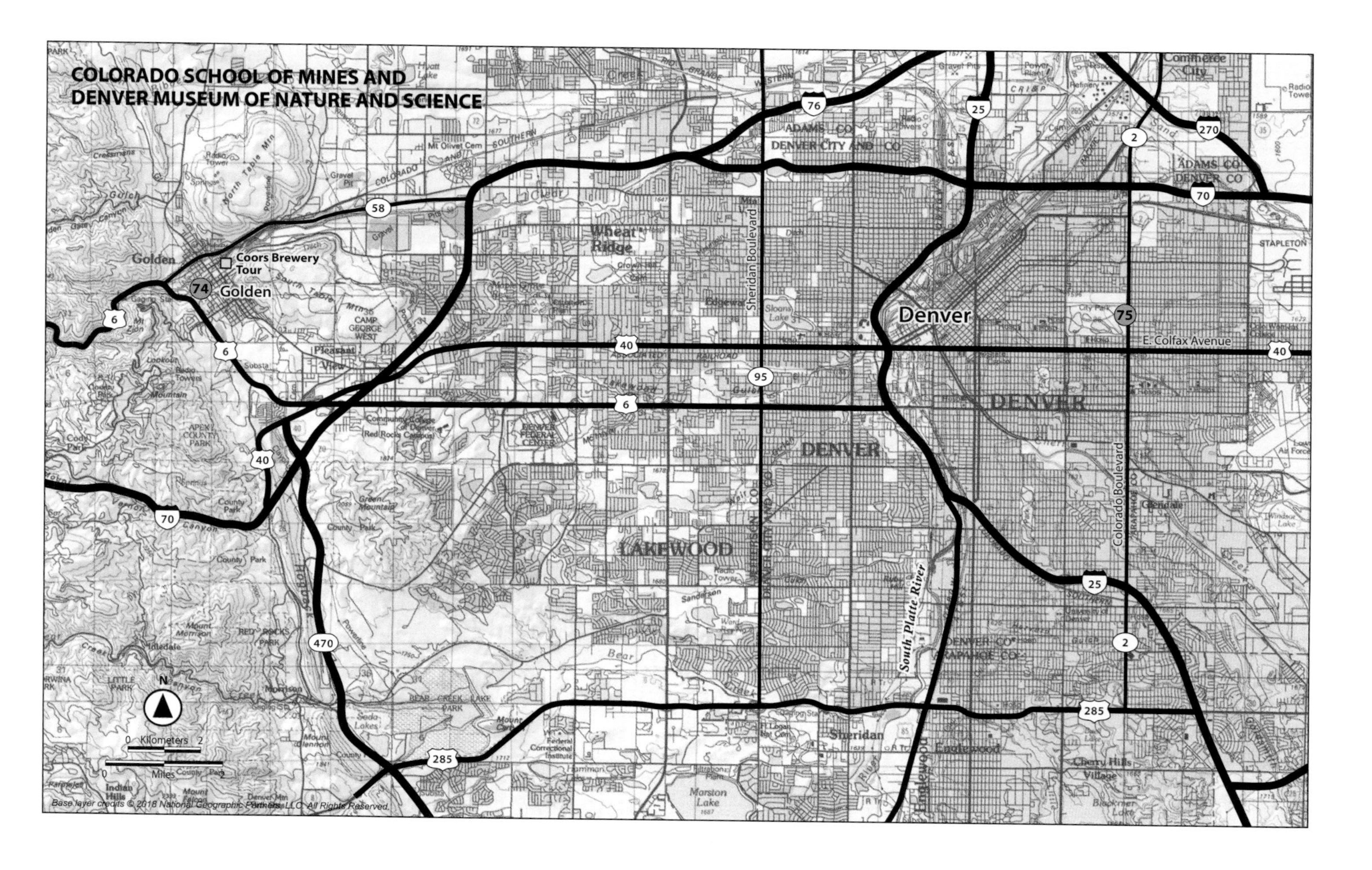

COLORADO SCHOOL OF MINES AND
DENVER MUSEUM OF NATURE AND SCIENCE
Coors Brewery Tour
74
Golden
75
Denver
E. Colfax Avenue
Sheridan Boulevard
Colorado Boulevard
DENVER
LAKEWOOD
Wheat Ridge
Sheridan
Englewood
Commerce City
STAPLETON
South Platte River
Glendale
Cherry Hills Village
76
25
270
2
70
58
6
40
95
470
285
N
0 Kilometers 2
0 Miles 2
Base layer credits © 2018 National Geographic Partners, LLC. All Rights Reserved.

You can find many of these great titles for your personal library at the museum's gift shop.

and the steady donations that continue to boost their collection. Of particular interest to gold prospectors is the Allison-Boettcher Collection, with spectacular leaf gold, crystalline gold, and wire gold from Colorado districts such as Telluride, Leadville, Central City, and Breckenridge. He told me that some of the most spectacular specimens in the collection were almost melted down for their gold value, but they ended up here. Dr. Geller was also proud of the pan full of Pennsylvania Mountain nuggets, specimens of pure native silver, and various telluride ores. Many of the districts captured in this book are represented in the displays.

The museum offers multiple teaching kits for rocks and minerals, fossils, and geologic concepts that it distributes throughout the school year to teachers. School tours are common, so the museum can seem quite busy at times. Another service is what I call "Stump the Geologist." They are happy to help with identifying a mystery rock or mineral that you bring in. It's a good idea to call ahead or check their website to make sure you get a knowledgeable person to assist you. Also, it's almost always better to be able to disclose where the specimen came from.

The museum is open year-round, except for major holidays. Winter break can also close the museum, so know that in advance. The museum closes at 3:45 p.m. each day, with limited hours Sunday. The good news is that admission is free. Visit www.mines.edu/Geology_Museum for more information.

75. Denver Museum of Nature & Science

Land type: Museum
County: Denver
Elevation: 5,315 feet
GPS: 39.74748, -104.94164
Best season: Any
Land manager: City of Denver
Material: Exhibits
Tools: Camera
Vehicle: Any
Special attractions: Western Museum of Mining & Industry in Colorado Springs
Accommodations: Full services in Denver
Finding the site: From I-70 eastbound, take exit 276B and stay right to merge onto CO 2/Colorado Boulevard. Drive 2.1 miles south and turn right onto Montview Boulevard to start looking for parking. The street address is 2001 Colorado Blvd.

Prospecting

The Denver Museum of Nature & Science originated in 1868, when Edwin Carter began his scientific study of the birds and mammals of the Rocky Mountains from a cabin in Breckenridge. In 1892 Carter sold his entire collection for $10,000 and it moved to Denver, where it joined a collection of crystallized gold. Opened to the public on July 1, 1908, the first president of the board, John F. Campion, declared, "A museum of natural history is never finished."

This is the crown jewel of Colorado museums, with more floor space, better specimens, and more family fun than any other destination. For gold panners, the top attraction by far is the Coors Mineral Hall, which boasts these must-see specimens:

1. The "Turtle Nugget" from the Bulger Basin Placer on top of Pennsylvania Mountain. Discovered in 1990, it weighs 7.8 ounces and is shaped like a turtle, with the head sticking out of the shell.

2. Colorado's largest gold nugget, also from the Pennsylvania Mountain diggings. It weights 12 troy ounces, or 1 pound, and was recovered in 1937.
3. The Campion Collection of wire gold from Farncomb Hill in Breckinridge. John F. Campion became rich mining in Leadville, and he used his money to purchase several mines at Farncomb Hill in 1894. He was an avid collector and paid the miners to bring him the best specimens. Campion helped found the Denver Museum in 1900, and his collection has been on display here since 1908.
4. "Tom's Baby," a 13.5-pound mass of wire gold discovered by Tom Groves and Harry Lytton in July 1887 at Farncomb Hill. According to author Heather Jarvis, the big mass of gold was shown around Breckenridge, weighed, assayed, cleaned up, and turned over to the mine owner, a man named Ward. The specimen was prepared for shipment to Denver via train, and it promptly disappeared. In 1972 a local writer in Breckenridge named Mark Fiester culminated years of research by tracking down the wooden box, marked "dinosaur bones," in a bank vault owned by the museum. Ward's daughter, who saw Tom's Baby as a child and was then in her 80s, was able to verify the specimen, which weighs only 10.5 pounds. The missing 3 pounds are probably also part of the Campion display.
5. The Summitville Boulder, whose discovery is covered in Site 31, weighs 114 pounds and contains an estimated 316 troy ounces of gold.

In addition to all that, there are exhibits and displays covering Colorado silver mining, Leadville, Cripple Creek, Gregory Gulch, Cache Creek, Cherry Creek, Russell Gulch, Georgetown, and more. Pay particular attention to the silver exhibit, since you don't want to repeat the mistake made by the early gold prospectors in Colorado, who overlooked the state's silver riches in their single-minded quest for gold. There are over 20,000 mineral specimens in the museum's collection, with only a fraction on display.

Other ore mineral exhibits to drool over include galena, sphalerite, pyrite, chalcopyrite, and many more; hopefully you can get your eye tuned for expeditions out to the tailings piles included in this book. Finally, there are beautiful mineral specimens of topaz, amazonite, aquamarine, and, of course, rhodochrosite. The famed Alma King rhodochrosite is on display, along with a giant 6-foot slab of rhodochrosite with its blood-red crystals still in place.

One of the most spectacular specimens on display at the Denver Museum of Nature & Science is "Tom's Baby," a 10.5-pound mass of crystalline gold from Farncomb Hill, near Breckenridge. It was discovered in the early days of the Breckenridge Camp, then "lost" in a box marked "dinosaur bones" in the Denver Museum vaults.

The Summitville Boulder, a 135-pound chunk of rock with hundreds of ounces of gold inside it, lay undiscovered in the parking lot at the Summitville Mine before a sharp-eyed truck driver spotted it.

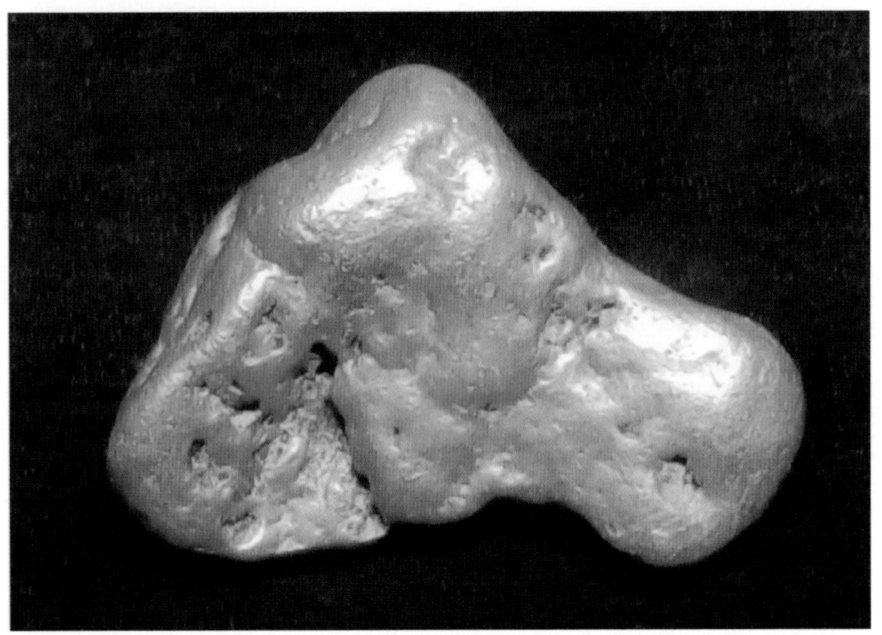

The biggest nugget discovered in Colorado was this 1-pound (12 troy ounces) monster from atop Pennsylvania Mountain near Alma.

The museum has additional science exhibits featuring Colorado dinosaurs and other fossils, partly thanks to a bulldozer operator that discovered ice age mammals at Snowmass Village. With exhibits on archaeology, a Discovery Zone, and numerous other attractions, you should plan to make a day of it, and leave with your own ideas on exploring the state. Check www.dmns.org for more information.

HONORABLE MENTIONS

The following locales did not make it into the book, but they're worth a visit if you're in the area.

Empire

North out of town on North Empire Road looked interesting, so is worth exploring, and it takes you to public lands

Fourmile Canyon

Access might be an issue. Try around 40.03659, -105.40517.

Gold Hill

Lots to explore around here at 40.06274, -105.40844, though there might be too much private land to dodge.

Hahn's Peak

The creeks around this locale, which is north of Eagle Lake off CO 139, are usually bone dry, but when running, they carry traces of gold. Hayden Hill itself was a major gold strike and is gated. You might try around 40.99874, -120.85102.

Horse Creek Placer

Never a big district production-wise, but probably worth checking at 39.24797, -105.22219 below Cheesman Reservoir.

Jamestown

The hills above Jamestown could be good, such as at 40.11210, -105.40302. Castle Creek at 40.11367, -105.35264 and Central Gulch at 40.13054, -105.36997 look good on paper.

Sugarloaf Drainage

Boulder Creek and the nearby area have good possibilities around 40.00447, -105.40548.

Vic's Gold Panning

On Highway 119 toward Black Hawk around MP 5, you can often see a sandwich board advertising gold-panning lessons. The turn is roughly 39.78329, -105.46352. Call ahead at (303) 582-0710.

Western Museum of Mining & Industry

225 North Gate Blvd., Colorado Springs. Use exit 156 north of Colorado Springs. Plenty of exhibits, indoor panning station, excellent gift shop, and more. Check www.wmmi.org for more details.

APPENDIX A: MODERN TOOLS

In my companion book, *The Modern Rockhounding and Prospecting Handbook*, available through FalconGuides, you'll learn the basics of reading geology maps, understanding symbols, figuring out basic geology, and more. In this section, you'll find overview information about various types of machines and devices that are out there.

One of the great joys of this hobby is sitting around the campfire after a long day and playing with your concentrates. (Second to that joy is doing the same thing in your garage, but campfires are better.) I like to bring buckets and a panning tub on my trips so that I can set up an area where I can screen and clean up the black sands and inspect the tailings with a hand lens for interesting rocks and minerals. At times I've had to dump the beer cans out of a cooler and use that for a panning station, however.

Buckets, Shovels, Picks, and Screens

I've met quite a few prospectors who enjoy bringing home a nice bucket of concentrates and then pan in the warmth of their garage. They don't like stretching out into a creek or river, trying to find a deep spot or risking a bad back, and they don't like getting their clothes and shoes wet when they inevitably slip. At home you can use warm water, stay out of the wind, put your panning tub up on a bench, add a drop of dispersant such as Jet-Dry to cut down on surface tension, and bring in plenty of light. So the first tools you need out there are buckets, a shovel, a pick for crevicing, and a screen.

Truthfully, screens are optional but a good idea, so you aren't bringing home a bunch of big rocks that you don't really need. Since the odds are billions and billions to one that you'll ever screen a big nugget, you might as well use screens to improve the quality of your pay dirt—whether you pan it out in the field or bring it home. Having said all that, fine gold can stick to rocks, so wash them off if you have the patience, with a scrub brush, old toothbrush, or similar tool. You'll know from the gold you're recovering whether a screen could catch big gold at your locale.

Shovels, buckets, picks, and screens do not require a dredging permit, and they're quiet. Your goal is to dig a big hole, because the deeper you go, the bigger the gold in 99 percent of the deposits. So it's up to you whether you bring the same shovel that you use in the garden or get a shorter shovel with a nice

You can never have enough buckets at a placer operation. Note the shorter, half-size buckets, which are easier to move around as the day wears on.

handle. On some trips you may use a trowel more than any other shovel, especially if you are digging out crevices in bedrock. The more you dig, and the deeper you go, the better your chances, so these are your first tools to consider.

Gold Pans

Your main tool is the venerable gold pan. Some older folks skip this step and go straight to spiral pans, saving wear and tear on their back, shoulders, and arms, but every prospector should have a pan or two in their collection.

The only way to get good at gold prospecting is through practice—you have to pan at least a dozen pans before you start to feel comfortable with sliding the heavies around, breaking up the muck, and trusting the riffles. There are twenty or more modern pan designs now, with sharp riffles, double sets of riffles for rough and fine panning, and all kinds of interesting shapes and additions. My personal favorite right now is a 14-inch octagonal pan with two sets of riffles and a broad, flat bottom. This pan helps with avoiding the tendency to swirl material in a circular pattern, which only serves to spread the concentrates around on the bottom. What you want to do is continually push heavy material into the first riffle and lock it in there, so you can start to feel more nonchalant about everything else in the pan.

When you are first breaking down a big pan full of muck, you need to learn to shake the pan vigorously so that the heavy material at the bottom slides forward to that first notch. This is called "stratifying," and it only works

Close-up showing the two sets of riffles on the 14-inch Octapan.

when you get the muck entirely into solution so that it's behaving like a liquid. Once you get your material into suspension, gravity will pull things apart, with light material like clay and debris floating to the top and heavy material like gold, garnets, and black sands falling to the bottom. So you want to do less swirling as you get closer to the end, and more side-to-side action. Pack that first riffle, and trust science.

The great thing about panning is that it's easy, and it's almost never banned. Even Wild and Scenic Rivers are open to panning below the high-water line. National parks and monuments are not open, however. State parks also tend to restrict panning, but county parks and USFS and BLM lands—even campgrounds—are usually open, and campgrounds are rarely claimed. So it's the first skill you need to gain as a prospector.

Snuffer Bottles

Once you've panned down a pan to where you can see what you have, you need some way to save the sample. You can use a wide-mouthed glass jar and just dump your concentrates in, although it's good to have a second pan underneath the jar. Most modern prospectors use a "snuffer bottle," which is a small plastic bottle with a removable straw. Once you're good at it, you can slurp up your black sands quickly, and there are two modes—without the

The more sample bottles you bring, the more concentrates you can clean up back at camp.

straw, for bulk recovery, and with the straw, for finer work. Once you get good with a snuffer bottle, you can make your specimen cleaner by, for example, picking up just the black sands and leaving the gold—if you have plenty of gold. You can easily fill up the small glass specimen jars by removing the straw and emptying the contents of the snuffer bottle; the heavies will concentrate in the tip. Give it a swirl when you're down to the final material.

Some folks use an adapted syringe to suck up their concentrates, then dump them into a specimen jar. I've seen someone use a turkey baster, too, then dump the concentrates into a finishing pan or a jar. When touring lots of places, I'll often just use one snuffer per location, and then figure it all out later. That's where good labeling comes in, and taking notes.

Sluices

The next step up from panning is the sluice, a modern adaptation of an old standby in the goldfields. Sluices are tricky because you have to get both the angle and the water flow correct. If you don't have enough water going in, you won't clean your gravels. If you have too much water, you can blow even heavies out the end. If you have the water and the angle adjusted correctly, then you're set. You can bring buckets of pay dirt to the sluice and wash a lot of material, and you will have a continual source of clean water, unlike

This sluice is set up with the right angle and the right water flow to process material.
PHOTO COURTESY OF DIRK WILLIAMS.

working with panning machines at camp. Sluices usually have three parts: the shell, a set of removable riffles, and a thick artificial turf or carpeting under the riffles. The heavies build up behind the riffles and in the carpeting, so you clean up by emptying material into a bucket and washing out the carpet.

When you have the sluice set up correctly, you don't need to clean up very often. But many crews like to clean up once or twice a day or more, just to make sure they don't lose anything. It's a good idea when you're first starting out to check frequently and compare against what you get with hand-panned samples. That way you can check your recovery. It's also very wise to grab a pan of material from right below the end of the sluice on occasion and pan that—just to see if your system is running properly.

Sluicing is a big step up in making concentrates, and you can move a lot of material—five or ten times what you can manually pan. At this point you are graduating from prospector to miner, and your goal is to keep feeding your sluice, keep moving material, and keep conducting cleanups. The down-side is that many areas are only open to panning, and a sluice is not appropri-ate. But in many areas where machines with pumps are not allowed, a sluice is still acceptable.

Once you've starting buying contraptions such as sluices and panning machines, you can advance to other devices. Desert miners swear by dry-washers, which use air to blow lighter material off the gold. There are books to help you get started with that kind of operation, which is beyond the scope of this book.

Battery-Powered Devices

There are two main kinds of panning machines: miniature sluices and spiral pans. Both work with a battery, usually 12-volt, and in a tub, although some versatile systems come with solar panels and work right in the stream. The miniature sluice usually has an upper deck and a lower deck, uses riffles or screens and some carpeting or plastic mini-riffles, and still traps a lot of black sands with the gold. Spiral pans work by washing off lighter material as the spiral brings material to the center, where it can either go through a hole to a little bucket in the back or stay in the middle in a little removable micro-bucket. The only real trick is getting the spray right, on a machine such as the Camel, or getting the tilt right, on a machine such as the Gold Magic. I like them both because they have their own tubs, and the only remaining challenge is access to clean water; you may have to empty and refill your tub every hour. We like to bring multiple buckets and pour the water in to let it settle, or in some places in the rainy season, we just drain rainwater from our blue tarp into buckets.

A recent invention is the Gold Cube, which is particularly effective at recovering fine gold. It's a nice addition for what the inventors call "21st Century Gold Prospecting." It uses new technology such as a G-Force separator, Vortex matting, and alternating modular trays. It runs on a 12-volt battery and is said to be capable of processing 1,000 pounds of screened material per hour. Go to www.goldcube.net to see some videos of the device in action.

Another new tool is the Gold Rush Nugget Bucket, which doesn't even require a battery or pump, although my friend Frank Higgins has been known to drain a battery or two using a converted aquarium pump to deliver water. This device uses a pair of screens and an ingenious funnel and bowl to trap all your heaviest concentrates. Check it out at www.goldrushnuggetbucket.com.

The beauty of these machines is that they are easy to set up and somewhat quiet, and if you have a picnic table back at camp, you can gather concentrates all day and pan at night by lantern. Panning machines work with rich concentrates you make from a dredge, sluice, trommel, or high-banker, or they'll work with the black sands you collected from the ten to twenty pans you processed that day. Some machines will even take raw pay dirt, but it's always better to screen the material so you're not dealing with rocks and pebbles.

Panning machines can alleviate some of the wear and tear on an "old dog" like my uncle, Doug Romaine.

With increasing restrictions on gas-powered dredging and high-banking, these battery-powered devices are becoming a popular option.

Another new invention is the hand dredge. There are several variations, such as using a bucket or a long piece of PVC pipe. The advantage is that these are hand tools, and never banned from public areas such as Cherry Creek. It takes a lot of patience to supply the power you would expect from a motor, as you are basically "slurping" up material by using your arms to pump back and forth. The payoff is that you can suck up material from around rocks and boulders right from the creek bed, and in underwater cracks and crevices. Once you get the motion down, you can pull up a lot of good material without the noise of a motorized dredge.

Metal Detectors and Pinpointers

Using a metal detector or a small, hand-size pinpointer takes patience and mechanical aptitude. You need to be willing to invest a few hours, at minimum, into swinging your detector before you can expect to be good at it. You need to dig all your targets, too, when starting out, so that you can tune your ear to what your detector is trying to tell you. The best detectors give you a reading so that you can get an idea of what your target might be, but

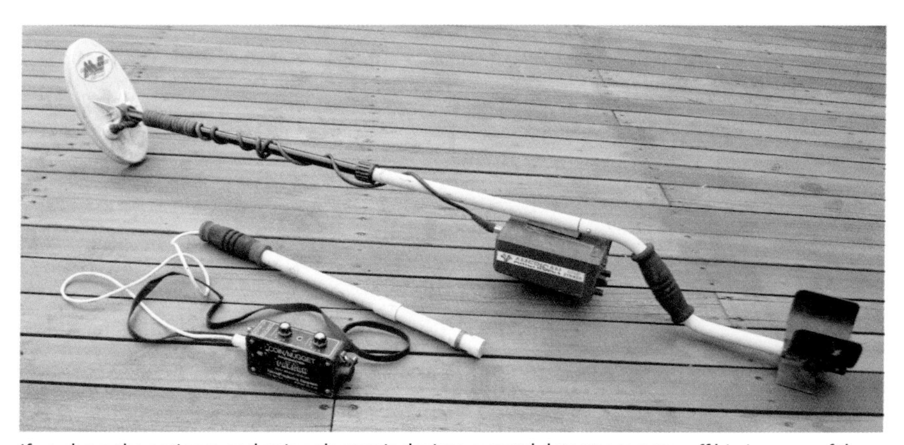

If you have the patience and enjoy electronic devices, a metal detector can pay off big in some of the large tailings fields found in the state of Colorado. Shown here is a small Falcon Gold Tracker pinpointer and an older Minelab device.

when starting out, you need to dig it up to verify that a nickel reading is actually a nickel. Some hot rocks will confuse your machine, and the basalt found all over Colorado can be a confounding rock in particular. In many places, you'll run into massive dredge fields with tailings piles that cover many, many acres. These are great places to hunt for gold nuggets, and you'll probably be removing a lot of nails, wires, screws, and scrap metal, too. The payoff can be enormous, and detecting can be a relaxing way to spend a day.

APPENDIX B: REFERENCE LIST AND FURTHER READING

I've been collecting books on Colorado mining history and gold prospecting for decades. Here are some of my favorite titles, with a few annotations to clarify their importance. I also listed some titles for further reading.

Bird, Allan G. 1999. Silverton *Gold: The Story of Colorado's Largest Gold Mine.* Self-published. Charming stories of the early years at Silverton.

Brown, Robert L. 2015. *Colorado Ghost Towns, Past and Present.* Caldwell, ID: Caxton Press.

Bureau of Land Management, 2016, Cache Creek Placer Area Fee Proposal, https://www.blm.gov/sites/blm.gov/files/uploads/Get%20Involved-RAC-Colorado-FR-8.18.16-Cache%20Creek%20Presentation.pdf.

Great stories about some of the more important, or notorious, old camps.

Burbank, W. S. 1932. *Geology and Ore Deposits of the Bonanza District.* USGS Professional Paper 169. Famed geologist spent ten months in the field at Bonanza and produced an epic report.

Cappa, James Al, and Paul J. Bartos, 2007. *Geology and Mineral Resources of Lake County, Colorado.* Colorado Department of Natural Resources, Resource Series 42.

Chenowith, William L. 1980. "Uranium in Colorado." In *Colorado Geology.* Denver, CO: Rocky Mountain Association of Geologists.

Chronic, Halka. 1980. *Roadside Geology of Colorado.* Missoula, MT: Mountain Press. Great helping hand for understanding the layer cake that is Colorado geology. After a while, you'll be able to spot the Morrison Formation from miles away.

Colorado Historical Society. 1986. *The Georgetown Loop: A Capsule History and Guide.* Denver, CO: Colorado Historical Society. Well-written, inexpensive little booklet on the history of Georgetown and its railroads.

Colorado State Planning Commission. 1936. *Unpublished Engineering and Geological Reports on Mineral Resources of Colorado.* Denver, CO: Colorado State Planning Commission.

Crawford, R. D. 1910. *A Preliminary Report on the Geology of the Monarch Mining District, Chaffee County, Colorado.* Colorado Geological Survey Bulletin 1.

Crawford, R. D., and Russell Gibson. 1925. *Geology and Ore Deposits of the Red Cliff District, Colorado.* Colorado Geological Survey Bulletin 30.

Dallas, Sandra. 1985. *Colorado Ghost Towns and Mining Camps.* Norman: University of Oklahoma Press. Excellent encyclopedia of both well-known and little-known camps. Good histories, too.

Dings, M. G., and C. S. Robinson, 1957. *Geology and Ore Deposits of the Garfield Quadrangle.* U.S. Geological Survey Professional Paper 289.

Downing, Charles V. 1993. *A Rockhound's Guide to Creede, Colorado.* Creede, CO: Tommyknocker. Good information if you're spending a lot of time at Creede. Available at the Railroad Depot museum.

Eckel, Edwin B., J. S. Williams, and F. W. Galbraith. 1949. *Geology and Ore Deposits of the La Plata District, Colorado.* U.S. Geological Survey Professional Paper 219.

Gold Prospectors Association of America. *GPAA Claims Guide.* Temecula, CA: GPAA. Best value for prospectors who cover multiple states and want to explore throughout North America. Includes multiple claims in the different regions of Colorado. Note that just owning the book doesn't make you a club member. The club stopped putting the year in their guide, so it's hard to know if you have the most current book.

Hague, Rick. 2016. "History: How Breckenridge Got It's Name," *Summit Daily.* https://www.summitdaily.com/explore-summit/history/history-how-breckenridge-got-its-name/.

Hansen, Wallace R. 1965. *The Black Canyon of the Gunnison, Today and Yesterday.* U.S. Geological Survey Bulletin 1191. The Gunnison wasn't a big mining district, but there is good information about the geology here.

Henderson, Charles W. 1926. *Mining in Colorado: A History of Discovery, Development, and Production.* U.S. Geological Survey Professional Paper 138.

Jackson, Donald Dale. 1980. *Gold Dust.* Lincoln: University of Nebraska Press. Easy-to-read overview of the western gold rushes.

Jarvis, Heather. 2015. "Discovering Tom's Baby, the Largest Piece of Gold Ever Found in Colorado," *Summit Daily.* https://www.summitdaily.com/news/discovering-toms-baby-the-largest-piece-of-gold-ever-found-in-colorado.

Jessen, Kenneth. 2012. "Snowstorm Dredge Near Fairplay the Only One Left in Colorado," *Loveland Reporter-Herald.* http://www.reporterherald.com/ci_22011861/snowstorm-dredge-near-fairplay-only-one-left-colorado.

Jessen, Kenneth. 2016. *Ghost Towns Colorado Style.* Vol. 1, *Northern Region.* Loveland, CO: J. V. Publications. A bit spendy, but good information about the old camps. Four separate volumes cover Colorado's different regions.

Johnson, Robert Neil. 1971. *Gold Diggers Atlas.* Susanville, CA: Cy Johnson & Son. An old favorite showing areas to explore. Very general information, not updated in decades, with questionable accuracy around roads, but includes historic places and won't let you down as far as finding historic districts.

Kappele, William. 2017. *Rockhounding Colorado: A Guide to the State's Best Rockhounding Sites.* Guilford, CT: FalconGuides. Updated by Gary and Sally Warren, with current information about rockhounding, especially on old mining dumps.

Kaysing, Bill. 1990. *Great Hot Springs of the West.* Santa Barbara, CA: Capra Press. Still relevant if looking for hot springs, both developed and off the beaten track.

Koschmann, A. H. 1949. *Structural Control of the Gold Deposits of the Cripple Creek District, Teller County, Colorado.* U.S. Geological Survey Bulletin 955-B.

Koschmann, A. H., and M. H. Bergendahl. 1968. *Principal Gold-Producing Districts of the United States.* U.S. Geological Survey Professional Paper 610. An excellent overview of the top gold-producing districts in the United States. Includes production figures up to 1968. Very general as far as placer information, and relies on old township/range location information. Excellent research guide for additional USGS reports when looking up specific information about a district. Great to have in PDF format.

Lindgren, Waldemar, and F. L. Ransome, 1906. *Geology and Gold Deposits of the Cripple Creek District, Colorado.* U.S. Geological Survey Professional Paper 54.

Lovering, T. S., and E.N. Goddard, 1950. *Geology and Ore Deposits of the Front Range, Colorado.* U.S. Geological Survey Professional Paper 223.

McKnight, Edwin Thor. 1974. *Geology and Ore Deposits of the Rico District, Colorado.* U.S. Geological Survey Professional Paper 723. Good resource for exploring the old mines and tailings piles around Rico.

Mitchell, James R. 1997. *Gem Trails of Colorado.* Baldwin Park, CA: Gem Guides Book Co. Old and dated, but gives a good idea of some of the tailings piles at old mines in the better districts.

———. 2007. *Gem Trails of Colorado.* 2nd ed. Baldwin Park, CA: Gem Guides Book Co. Updated, and even better.

Monaco, James M., and Jeanette H. Monaco. 1997. *Fee Mining Adventures and Rockhounding Expeditions.* Baldwin Park, CA: Gem Guides Book Co. Getting a little dated.

Muilenberg, Garrett A. 1925. *Geology of the Tarryall District, Park County, Colorado.* Colorado Geological Survey Bulletin 31.

Parker, Ben H. 1968. "Placer Gold in Southwestern Colorado." In *San Juan, San Miguel, La Plata Region (New Mexico and Colorado),* edited by J. W. Shomaker, 168–84. New Mexico Geological Society 19th Annual Fall Field Conference Guidebook. Another classic, and incredibly valuable for the southwest region. Just a little dated, however.

———. 2009. *Gold Panning and Placering in Colorado: How and Where.* 2nd ed. Denver, CO: Colorado Geological Survey, Department of Natural Resources. Essential guide for anyone researching placer opportunities. This book started as Parker's doctoral thesis at the Colorado School of Mines. Parker spent years in the many districts of Colorado, and his knowledge of alluvial deposits, colluvial deposits, glacial till, terraces, benches, skim-bar accretion deposits, and more has guided professional and amateur gold prospectors since this book first came out. The first edition was already a classic, but the second edition is in color, is better organized, has more general information for beginners, and is an even better read.

Patton, Horace B. 1917. *Geology and Ore Deposits of the Platoro-Summitville Mining District, Colorado.* Colorado Geological Survey Bulletin 13.

Patton, Horace B., Arthur J. Hoskin, and G. Montague Butler. 1912. *Geology and Ore Deposits of the Alma District, Park County, Colorado.* Colorado State Geological Survey Bulletin 3.

Patton, Horace B., Charles E. Smith, G. Montague Butler, and Arthur J. Hoskin. 1910. *Geology of the Grayback Mining District, Costilla County, Colorado.* Colorado State Geological Survey Bulletin 2.

Paulson, Don. 2015. *Mines, Miners, and Much More: A Guide to Historic Mining in Colorado's San Juan Triangle.* Chicago: Twain Publishers. Good history and excellent information about road trips around Ouray, Silverton, and Telluride. Meticulously researched, with a good bibliography containing references to articles in long-lost newspapers.

Ransome, Frederick Leslie. 1911. *Geology and Ore Deposits of the Breckenridge District, Colorado.* U.S. Geological Survey Professional Paper 75.

Ridge, John D. 1968. *Ore Deposits in the United States: 1933–1967.* New York: American Institute of Mining, Metallurgical, and Petroleum Engineers. Part V covers the Colorado Rockies.

Romaine, Garret. 2014. *The Modern Rockhounding and Prospecting Handbook.* Guilford, CT: FalconGuides. Good overview for prospecting and field geology, with details about reading geology maps, identifying key rocks and minerals, planning successful trips, and more.

———. 2014. *Rocks, Gems, and Minerals of the Rocky Mountains.* Guilford, CT: FalconGuides. Pocket-sized for easy reference in the field.

Town of Breckenridge, *Dredge Mining and the Early 1900s,* http://www.townofbreckenridge.com/live/heritage-history/town-history/dredge-mining-and-the-early-1900s.

U.S. Geological Survey. 1908. *Mineral Resources of the United States—Calendar Year 1907.* Washington, DC: U.S. Government Printing Office. Any of these reports are good to have for overview information, even though out-of-date.

Vhay, John S. 1962. *Geology and Mineral Deposits of the Area South of Telluride, Colorado.* U.S. Geological Survey Bulletin 1112-G. An example of the great contributions geologists have made to understanding the complex geology and mineralogy of Colorado's gold and silver districts.

Voynick, Stephen M. 1984. *Leadville: A Miner's Epic.* Missoula, MT: Mountain Press. Excellent resource, with key information about Leadville mining.

———. *Colorado Rockhounding.* 1994. Missoula, MT: Mountain Press. Must-have from one of the giants of Colorado gold prospecting, mineral collecting, and fossil hunting. Packs in over 350 sites, and as far as gold panning, will steer you away from certain sites, too. That can save you a long trip!

Voynick, Stephen M. "The Summitville 141-Pound Gold Boulder," *Pagosa Springs Journal.* Posted March 31, 2014. http://pagosasprings.com/the-summitville-141-oz-gold-boulder.

Voynick, Stephen M. December 2016 issue of *Rock and Gem* magazine. Lake City, Colorado, pp. 38-41.

Weston, William. 2006. *Descriptive Pamphlet of Some of the Principal Mines and Prospects of Ouray & San Miguel Counties 1882–3 in the San Juan Gold and Silver Region.* Montrose, CO: Western Reflections. More old history from one of the earliest geologists in the region. He may call himself simply a "miner," but he was an assayer, mining engineer, and promoter. He has a charming, engaging writing style, too.

Worcester, P. G. 1919. *Molybdenum Deposits of Colorado.* Colorado Geological Survey Bulletin 14.

Young, H. Court. 2007. *The Orphan Boy: A Love Affair With Mining.* Beautifully told story by the son of Herbert Thompson Young about his father's efforts in re-opening the Orphan Boy Mine in the Montezuma Mining District near Alma.

APPENDIX C: WEBSITES

I've been tracking gold-prospecting links on the Internet since 1996 and wrote a regular column for *Gold Prospector* magazine titled "Mining the Internet." If there's one thing you can count on, it's the fact that things change quickly in cyberspace. These are great places to begin your online research, but your web sleuthing could turn up more good information.

Ancestry.com
The resources on this site can help you find information about old ghost towns and mining camps. For example, this search pulled up a long list of old camps for Gunnison County: http://rootsweb.ancestry.com/~cogunnis/cities.htm.

Bureau of Land Management (BLM)
www.blm.gov/colorado
Link to local offices for Colorado from here.

Colorado Department of Transportation
www.codot.gov/travel
Road conditions, travel alerts, and, best of all, online cameras.

Colorado Geological Survey–Mineral Resources
http://coloradogeologicalsurvey.org/mineral-resources
Access historic mining district information and find loads of old reports.

Colorado Tourism
www.colorado.com
Good starting point, with free visitor guides, maps, videos, articles, and inspiration.

The Diggings
https://thediggings.com/usa/colorado
Helpful research site. You can get a good idea about activity out in the mountains by checking to see the changes in the number of staked claims, for example.

Earth Point
www.earthpoint.us/TopoMap.aspx
Google Earth overlay that brings in topo maps.

Finding Gold in Colorado
http://findinggoldincolorado.com
Kevin Singel operates a great blog and constantly updates it. The companion to it is his Facebook page under the same name.

Fishing Locations
http://cpw.state.co.us/placestogo/Pages/FishingStatewideMaps.aspx
Many times, if the anglers have access, you will too.

Ghost Towns
www.ghosttowns.com
County-by-county interactive map with good information about many mining camps and old towns. Lots of pictures.

Gold Cube
https://goldcube.net
Super-efficient battery-powered device perfect for Colorado's fine gold. Excellent blog, too.

Gold Panning Championships
www.worldgoldpanningassociation.com
Speed-panning at its finest.

Gold Prospecting Online
www.goldprospectingonline.com
Good site for general gold-mining information, helpful info for beginners, and sells equipment.

Gold Rush Expeditions
https://goldrushexpeditions.com/state/colorado
Claims for sale.

Gold Rush Nugget Bucket
www.goldrushnuggetbucket.com
Check out the videos for this inexpensive new device.

Gold Rush Nuggets
www.goldrushnuggets.com
Good articles, nuggets for sale, and more.

Gold Strike Prospecting
www.goldstrikecoloradoprospecting.blogspot.com
Excellent resource about panning locales.

Gold Unlimited
https://goldunlimited.org
Good spot for information, especially about Denver-area sites.

Goldstrike Adventures
http://goldstrikeadventures.com
Learn from the best with a tailored trip and guide. They sell bags of pay dirt, too.

Google Earth
www.google.com/earth
One of my favorite programs for tracking where I've been. I have a GPS device that I can synch with Google Earth and import my waypoints and tracks, which sometimes shows how close I got to a place I had my heart set on visiting.

Grubstaker
www.grubstaker.com
Plenty of information on claims currently for sale in Colorado.

Hiking and Biking Trails
www.alltrails.com
Can be good information about hiking to mining districts.

Hooked on Gold Prospectors Blog
http://hookedongold.blogspot.com
Fun reads; good info. Not updated regularly.

Mindat
www.mindat.org
Primarily a site for mineral collectors, this extensive database has good information for gold prospectors, too. You can figure out what minerals you recovered from a tailings pile, for example.

Mine Finder
https://mrdata.usgs.gov/mrds/find-mrds.php
Mineral Resources Data System mine finder.

MineCache
www.minecache.com
This overlay places icons for gold mines and major prospects into Google Earth. It's a bit spendy, and it's not verified in the field, but it gives you a good idea of what major mine is nearby.

Mineral Resources Data System (MRDS)
http://mrdata.usgs.gov/mrds
An excellent source of information packed by the US Geological Survey, with detailed information about the geology of significant mines in an area. You can export information to build your own spreadsheet as well. One downside is that the general map includes sand and gravel operations, so it helps to sort by commodity before starting. The report also often describes which particular references include more information, and with more and more of these coming online as Acrobat files, your online researching promises to yield better and better results.

MyTopo
www.mytopo.com/maps/index.cfm
Zoom in on your area and pull up the most recent topo map. This is quick and easy, but sometimes you want those historical maps, and in that case, you'd use TopoView.

Price of Gold
http://onlygold.com/Info/Historical-Gold-Prices.asp
Sometimes the old literature will only tell you how much money was extracted from a mine, not how many ounces of gold. Use this site to help with your conversion. As recently as 1965, gold sold for only about $35 per ounce. In 1848, the price was $20.67.

Prospectors Paradise
www.prospectorsparadise.com
General info about gold mining, and an active forum for asking questions.

Rare Gold Nuggets
http://raregoldnuggets.com
Good overviews and history.

Recreation.gov
Discover and reserve camping, lodging, permits, tours, and more at America's parks, forests, monuments, and other public lands.

Rocky Mountain Association of Geologists
www.rmag.org
Committed to geology education in the Rocky Mountain states.

Scenic Byways
www.codot.gov/travel/scenic-byways
Good information for official routes through the passes. Grab the virtual guide, use the map, and plan your trip.

The GeoZone
www.thegeozone.com/treasure/colorado/index.jsp
Especially good for lost treasure information.

Tom Ashworth's Gold Fever Prospecting (was Prospector Cache)
www.goldfeverprospecting.com/cogolo.html
Excellent overview information about gold in Colorado. Good for research, but not up-to-date.

TopoView
https://ngmdb.usgs.gov/topoview
Replacement for GeoCommunicator, with access to historic topo maps that show the old mines. It doesn't have the ease of use of the old system, but it has more power to access the old maps. Getting the filtering right takes some practice. Remember that 1:24,000 is the most detail.

TreasureNet
www.treasurenet.com
Great forums.

US Forest Service
www.fs.fed.us
Good starting point to find the local districts and forest managers for areas you plan to visit. Don't be bashful—give them a call. Use the site for information about flooding, road washouts and closures, fire restrictions, and campsites. If you plan to get away from it all and practice dispersed, primitive camping, you need these resources, including their maps.

US Geological Survey
www.usgs.gov
Learn the science behind gold prospecting.

US Mining
www.us-mining.com/colorado
Excellent resource for finding mines by county and commodity. They don't pinpoint the area with GPS coordinates, unfortunately.

Water Levels (USGS)

http://waterdata.usgs.gov

Good for determining how high the water is at a stream you desperately want to cross or sluice.

Western Mining History

http://westernmininghistory.com

Good background information about the major gold-mining districts in the West.

World Wide Association of Treasure Seekers

www.wwats.org

Well-known group in southwestern Colorado. Wayne "Nugget Brain" Peterson is a local legend among gold panners and metal detector enthusiasts.

APPENDIX D: CLUBS AND ORGANIZATIONS

I cannot stress enough the importance of joining a gold-prospecting club. Not only do you get access to valid claims, you also begin building a network of contacts with knowledge about prospecting equipment and regional hot spots. By regularly attending club meetings and annual shows, you'll learn what's new and interesting, and soon you'll be answering questions and giving back. Some clubs even have equipment you can borrow or rent, and some bring equipment to meetings so you can practice, clean up concentrates, etc.

The downside is that it would be nice to get more reciprocal benefits, especially for those of us who like to roam from state to state, or even from district to district. It is annoying to hear, "You're in the wrong club!"

Below are some links to get you started. I did my best to update everything and scoured the web for clubs I didn't know, but it's possible I missed some. More groups are relying on Facebook to host their web presence, so be sure to try some searches there as well.

National

Gold Miner's Headquarters
www.goldminershq.com/clubs/gold2.htm
Contains a list of clubs by state.

Gold Prospectors Association of America (GPAA)
www.goldprospectors.org
My personal favorite, with claims in all western states and a large, active reach. They have annual shows in several US cities and an excellent magazine that I've written for on and off since 1996. They have many, many local chapters; Colorado is no exception.

Prospecting Channel
www.prospectingchannel.com/indexClub.html
Good updated list for clubs both national and worldwide.

Regional

Colorado Gold Camp
www.coloradogoldcamp.com
Family-friendly Colorado prospecting club.

Colorado Prospector
www.coloradoprospector.com
A Colorado Prospector club membership grants you access to their research. Researched prospecting areas for club members currently total over 30,000.

Gold Prospectors of Colorado
www.gpoc.club
Colorado's premiere gold-prospecting club.

Western Colorado Chapter of Gold Prospectors Association of America
www.wcgpaa.org
The claims used by the club belong to the Gold Prospectors Association of America.

Western Mining Alliance
http://westernminingalliance.org
Large organization of independent miners.

ABOUT THE AUTHOR

Garret Romaine is an avid gold prospector, rockhound, and fossil collector with years of experience in the field. He is a longtime writer for *Gold Prospectors* magazine and is the author of *Gold Panning the Pacific Northwest; Gold Panning California; The Modern Rockhounding and Prospecting Handbook*; *Rocks, Gems, and Minerals of the Southwest*; *Rocks, Gems, and Minerals of the Rocky Mountains*; and *Rockhounding Idaho*, all from FalconGuides, as well as *Gem Trails of Washington* and *Gem Trails of Oregon*. Garret is the Secretary of the Board of Directors for the Rice Northwest Museum of Rocks and Minerals in Hillsboro, Oregon.